Accelerate Model Training with PyTorch 2.X

Build more accurate models by boosting the model training process

Maicon Melo Alves

Accelerate Model Training with PyTorch 2.X

Copyright © 2024 Packt Publishing

Group Product Manager: Niranjan Naikwadi
Publishing Product Manager: Sanjana Gupta
Book Project Manager: Kirti Pisat
Content Development Editor: Manikandan Kurup
Technical Editor: Seemanjay Ameriya
Copy Editor: Safis Editing
Proofreader: Safis Editing and Manikandan Kurup
Indexer: Hemangini Bari
Production Designer: Aparna Bhagat
Senior DevRel Marketing Coordinator: Vinishka Kalra

First published: April 2024

Production reference: 1050424

Published by Packt Publishing Ltd.
Grosvenor House
11 St Paul's Square
Birmingham
B3 1RB, UK.

ISBN 978-1-80512-010-0

www.packtpub.com

To my wife and best friend, Cristiane, for being my loving partner throughout our joint life journey. To my daughters, Giovana and Camila, for being my real treasure; I'm so proud of you. To my mom, Fatima, and brothers, Johny and Karoline, for being my safe harbor. Despite everything, I also dedicate this book to my (late) father Jorge.

– Maicon Melo Alves

Foreword

Accelerating model training is critical in the area of machine learning for several reasons. As datasets grow larger and models become more complex, training times can become prohibitively long, hindering research and development progress. This is where machine learning frameworks such as PyTorch come into play, providing tools and techniques to accelerate the training process.

PyTorch, with its flexibility, GPU acceleration, optimization techniques, and distributed training capabilities, plays a crucial role in this endeavor by enabling researchers and developers to iterate quickly, train complex models efficiently, and deploy solutions faster. By leveraging PyTorch's capabilities, practitioners can push the boundaries of what is possible in artificial intelligence and drive innovation across various domains.

Since learning all of these capabilities is not a straightforward task, this book is a great resource for all students, researchers, and professionals who intend to learn how to accelerate model training with the latest release of PyTorch in a smooth way.

This very didactic book starts by introducing how the training process works and what kind of modifications can be done at the application and environment layers to accelerate the training process.

Only after that, the following chapters describe methods to accelerate the training model, such as the Compile API, a novel capability launched in PyTorch 2.0 useful for compiling a model, and the use of specialized libraries such as OpenMP and IPEX to speed up the training process of our models even more.

It also describes the building of an efficient data pipeline to keep your GPU working at its peak for the entire training process, simplifying a model by reducing the number of parameters, and reducing the numerical precision adopted by the neural network to accelerate the training process and decrease the amount of memory needed to store the model.

Finally, this book also explains how to spread out the distributed training process to run on multiple CPUs and GPUs.

This book not only provides current and highly relevant content for the learning and updating of any professional working in the field of computing but also impresses with its extremely didactic presentation of the subject. You will certainly appreciate the quiz at the end of each chapter and the connection made between the chapters in the summary at the end of each chapter.

In all chapters, codes, and examples of use are presented. For all these reasons, I believe that the book could be successfully adopted by undergraduate and graduate courses as a support bibliography for them too.

Prof. Lúcia Maria de Assumpção Drummond

Titular professor at Fluminense Federal University, Brazil

Contributors

About the author

Dr. Maicon Melo Alves is a senior system analyst and academic professor who specializes in **High-Performance Computing** (**HPC**) systems. In the last five years, he has become interested in understanding how HPC systems have been used in AI applications. To better understand this topic, he completed an MBA in data science in 2021 at Pontifícia Universidade Católica of Rio de Janeiro (PUC-RIO). He has over 25 years of experience in IT infrastructure, and since 2006, he has worked with HPC systems at Petrobras, the Brazilian state energy company. He obtained his DSc degree in computer science from the **Fluminense Federal University** (**UFF**) in 2018 and has published three books and publications in international journals in HPC.

About the reviewer

Dimitra Charalampopoulou is a machine learning engineer with a background in technology consulting and a strong interest in AI and machine learning. She has led numerous large-scale digital transformation engineering projects for clients across the US and EMEA and has received various awards, including recognition for her start-up at the MIT Startup Competition. Additionally, she has been a speaker at two conferences in Europe on the topic of GenAI. As an advocate for women in tech, she is the founder and managing director of an NGO that promotes gender equality in tech and has taught programming classes to female students internationally.

Table of Contents

Part 1: Paving the Way

1

Deconstructing the Training Process   3

2

Training Models Faster   19

Part 2: Going Faster

3

Part 3: Going Distributed

11

Training with Multiple Machines   179

Preface

Hello there! I'm a system analyst and academic professor specializing in **High-Performance Computing (HPC)**. Yes, you read it right! I'm not a data scientist. So, you are probably wondering why on Earth I decided to write a book about machine learning. Don't worry; I will explain.

HPC systems comprise powerful computing resources tightly integrated to solve complex problems. The main goal of HPC is to employ resources, techniques, and methods to accelerate the execution of highly intensive computing tasks. Traditionally, HPC environments have been used to execute scientific applications from biology, physics, chemistry, and many other areas.

But this has changed in the past few years. Nowadays, HPC systems run tasks beyond scientific applications. In fact, the most prominent non-scientific workload executed in HPC environments is precisely the subject of this book: the building process of complex neural network models.

As a data scientist, you know better than anyone else how long it could take to train complex models and how many times you need to retrain the model to evaluate different scenarios. For this reason, the usage of HPC systems to accelerate **Artificial Intelligence (AI)** applications (not only for training but also for inference) is a growth-demanding area.

This close relationship between AI and HPC sparked my interest in diving into the fields of machine learning and AI. By doing this, I could better understand how HPC has been applied to accelerate these applications.

So, here we are. I wrote this book to share what I have learned about this topic. My mission here is to give you the necessary knowledge to train your model faster by employing optimization techniques and methods using single or multiple computing resources.

By accelerating the training process, you can concentrate on what really matters: building stunning models!

Who this book is for

This book is for intermediate-level data scientists, engineers, and developers who want to know how to use PyTorch to accelerate the training process of their machine learning models. Although they are not the primary audience for this material, system analysts responsible for administrating and providing infrastructure for AI workloads will also find valuable information in this book.

Basic knowledge of machine learning, PyTorch, and Python is required to get the most out of this material. However, there is no obligation to have a prior understanding of distributed computing, accelerators, or multicore processors.

What this book covers

Chapter 1, Deconstructing the Training Process, provides an overview of how the training process works under the hood, describing the training algorithm and covering the phases executed by this process. This chapter also explains how factors such as hyperparameters, operations, and neural network parameters impact the training process's computational burden.

Chapter 2, Training Models Faster, provides an overview of the possible approaches to accelerate the training process. This chapter discusses how to modify the application and environment layers of the software stack to reduce the training time. Moreover, it explains vertical and horizontal scalability as another option to improve performance by increasing the number of resources.

Chapter 3, Compiling the Model, provides an overview of the novel Compile API introduced on PyTorch 2.0. This chapter covers the differences between eager and graph modes and describes how to use the Compile API to accelerate the model-building process. This chapter also explains the compiling workflow and components involved in the compiling process.

Chapter 4, Using Specialized Libraries, provides an overview of the libraries used by PyTorch to execute specialized tasks. This chapter describes how to install and configure OpenMP to deal with multithreading and IPEX to optimize the training process on an Intel CPU.

Chapter 5, Building an Efficient Data Pipeline, provides an overview of how to build an efficient data pipeline to keep the GPU working as much as possible. Besides explaining the steps executed on the data pipeline, this chapter describes how to accelerate the data-loading process by optimizing GPU data transfer and increasing the number of workers on the data pipeline.

Chapter 6, Simplifying the Model, provides an overview of how to simplify a model by reducing the number of parameters of the neural network without sacrificing the model's quality. This chapter describes techniques used to reduce the model complexity, such as model pruning and compression, and explains how to use the Microsoft NNI toolkit to simplify a model easily.

Chapter 7, Adopting Mixed Precision, provides an overview of how to adopt a mixed precision strategy to burst the model training process without penalizing the model's accuracy. This chapter briefly explains numeric representation in computer systems and describes how to employ PyTorch's automatic mixed precision approach.

Chapter 8, Distributed Training at a Glance, provides an overview of the basic concepts of distributed training. This chapter presents the most adopted parallel strategies and describes the basic workflow to implement distributed training on PyTorch.

Chapter 9, Training with Multiple CPUs, provides an overview of how to code and execute distributed training in multiple CPUs on a single machine using a general approach and Intel oneCCL to optimize the execution on Intel platforms.

Chapter 10, Training with Multiple GPUs, provides an overview of how to code and execute distributed training in a multi-GPU environment on a single machine. This chapter presents the main characteristics of a multi-GPU environment and explains how to code and launch distributed training on multiple GPUs using NCCL, the default communication backend for NVIDIA GPUs.

Chapter 11, Training with Multiple Machines, provides an overview of how to code and execute distributed training in multiple GPUs on multiple machines. Besides an introductory explanation of computing clusters, this chapter shows how to code and launch distributed training among multiple machines using Open MPI as the launcher and NCCL as the communication backend.

To get the most out of this book

You will need to have an understanding of the basics of machine learning, PyTorch, and Python.

Software/hardware covered in the book	Operating system requirements
PyTorch 2.X	Windows, Linux, or macOS

If you are using the digital version of this book, we advise you to type the code yourself or access the code from the book's GitHub repository (a link is available in the next section). Doing so will help you avoid any potential errors related to the copying and pasting of code.

Download the example code files

You can download the example code files for this book from GitHub at `https://github.com/PacktPublishing/Accelerate-Model-Training-with-PyTorch-2.X`. If there's an update to the code, it will be updated in the GitHub repository.

We also have other code bundles from our rich catalog of books and videos available at `https://github.com/PacktPublishing/`. Check them out!

Conventions used

There are a number of text conventions used throughout this book.

`Code in text`: Indicates code words in text, database table names, folder names, filenames, file extensions, pathnames, dummy URLs, user input, and Twitter handles. Here is an example: "The `ipex.optimize` function returns an optimized version of the model."

A block of code is set as follows:

```
config_list = [{
    'op_types': ['Linear'],
    'exclude_op_names': ['layer4'],
    'sparse_ratio': 0.3
}]
```

When we wish to draw your attention to a particular part of a code block, the relevant lines or items are set in bold:

```
def forward(self, x):
    out = self.layer1(x)
    out = self.layer2(out)
    out = out.reshape(out.size(0), -1)
    out = self.fc1(out)
    out = self.fc2(out)
    return out
```

Any command-line input or output is written as follows:

```
maicon@packt:~$ nvidia-smi topo -p -i 0,1
Device 0 is connected to device 1 by way of multiple PCIe
```

Bold: Indicates a new term, an important word, or words that you see onscreen. For instance, words in menus or dialog boxes appear in **bold**. Here is an example: "**OpenMP** is a library used for parallelizing tasks by harnessing all the power of multicore processors by using the multithreading technique."

> **Tips or important notes**
> Appear like this.

Get in touch

Feedback from our readers is always welcome.

General feedback: If you have questions about any aspect of this book, email us at customercare@ packtpub.com and mention the book title in the subject of your message.

Errata: Although we have taken every care to ensure the accuracy of our content, mistakes do happen. If you have found a mistake in this book, we would be grateful if you would report this to us. Please visit www.packtpub.com/support/errata and fill in the form.

Piracy: If you come across any illegal copies of our works in any form on the internet, we would be grateful if you would provide us with the location address or website name. Please contact us at copyright@packt.com with a link to the material.

If you are interested in becoming an author: If there is a topic that you have expertise in and you are interested in either writing or contributing to a book, please visit authors.packtpub.com.

Share Your Thoughts

Once you've read *Accelerate Model Training with PyTorch 2.X*, we'd love to hear your thoughts! Scan the QR code below to go straight to the Amazon review page for this book and share your feedback.

https://packt.link/r/1-805-12010-7

Your review is important to us and the tech community and will help us make sure we're delivering excellent quality content.

Download a free PDF copy of this book

Thanks for purchasing this book!

Do you like to read on the go but are unable to carry your print books everywhere?

Is your eBook purchase not compatible with the device of your choice?

Don't worry, now with every Packt book you get a DRM-free PDF version of that book at no cost.

Read anywhere, any place, on any device. Search, copy, and paste code from your favorite technical books directly into your application.

The perks don't stop there, you can get exclusive access to discounts, newsletters, and great free content in your inbox daily

Follow these simple steps to get the benefits:

1. Scan the QR code or visit the link below

https://packt.link/free-ebook/978-1-80512-010-0

2. Submit your proof of purchase
3. That's it! We'll send your free PDF and other benefits to your email directly

Part 1:
Paving the Way

In this part, you will learn about performance optimization, before delving into the techniques, approaches, and strategies described throughout the book. First, you will learn about the aspects of the training process that make it so computationally heavy. After that, you will learn about the possible approaches to reduce the training time.

This part has the following chapters:

- *Chapter 1, Deconstructing the Training Process*
- *Chapter 2, Training Models Faster*

1
Deconstructing the Training Process

We already know that training neural network models takes a long time to finish. Otherwise, we would not be here discussing ways to run this process faster. But which characteristics make the building process of these models so computationally heavy? Why does the training step take so long? To answer these questions, we need to understand the computational burden of the training phase.

In this chapter, we will first remember how the training phase works under the hood. We will understand what makes the training process so computationally heavy.

Here is what you will learn as part of this first chapter:

- Remembering the training process
- Understanding the computational burden of the training phase
- Understanding the factors that influence training time

Technical requirements

You can find the complete code of the examples mentioned in this chapter in the book's GitHub repository at `https://github.com/PacktPublishing/Accelerate-Model-Training-with-PyTorch-2.X/blob/main`.

You can access your favorite environment to execute this notebook, such as Google Colab or Kaggle.

Remembering the training process

Before describing the computational burden imposed by neural network training, we must remember how this process works.

> **Important note**
>
> This section gives a very brief introduction to the training process. If you are totally unfamiliar with this topic, you should invest some time to understand this theme before moving to the following chapters. An excellent resource for learning this topic is the book entitled *Machine Learning with PyTorch and Scikit-Learn*, published by Packt and written by Sebastian Raschka, Yuxi (Hayden) Liu, and Vahid Mirjalili.

Basically speaking, neural networks learn from examples, similar to a child observing an adult. The learning process relies on feeding the neural network with pairs of input and output values so that the network catches the intrinsic relation between the input and output data. Such relationships can be interpreted as the knowledge obtained by the model. So, where a human sees a bunch of data, the neural network sees veiled knowledge.

This learning process depends on the dataset used to train a model.

Dataset

A **dataset** comprises a set of **data instances** related to some problem, scenario, event, or phenomenon. Each instance has features and target information corresponding to the input and output data. The concept of a dataset instance is similar to a registry in a table or relational database.

The dataset is usually split into two parts: training and testing sets. The training set is used to train the network, whereas the testing part is used to test the model against unseen data. Occasionally, we can also use another part to validate the model after each training iteration.

Let's look at Fashion-MNIST, a famous dataset that is commonly used to test and teach neural networks. This dataset comprises 70,000 labeled images of clothes and accessories such as dresses, shirts, and sandals belonging to 10 distinct classes or categories. The dataset is split into 60,000 instances for training and 10,000 instances for testing.

As shown in *Figure 1.1*, a single instance of this dataset comprises a 28 x 28 grayscale image and a label identifying its class. In the case of Fashion-MNIST, we have 70,000 instances, which is often referred to as the length of the dataset.

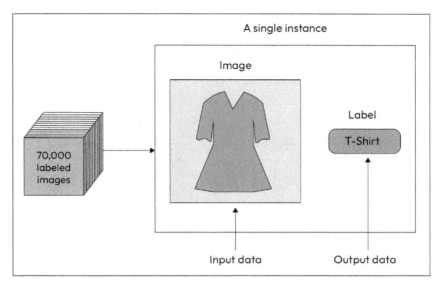

Figure 1.1 – Concept of a dataset instance

Besides the concept of dataset instance, we also have the concept of a **dataset sample**. A sample is defined as a group of instances, as shown in *Figure 1.2*. Usually, the training process executes on samples and not just on a single dataset instance. The reason why the training process takes samples instead of single instances is related to the way the training algorithm works. Don't worry about this topic; we will cover it in the following sections:

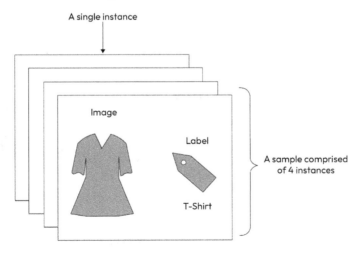

Figure 1.2 – Concept of a dataset sample

The number of instances in a sample is called the **batch size**. For example, if we divide the Fashion-MNIST training set into samples of a batch size equal to 32, we get 1,875 samples since this set has 60,000 instances.

The higher the batch size, the lower the number of samples in a training set, as pictorially described in *Figure 1.3*:

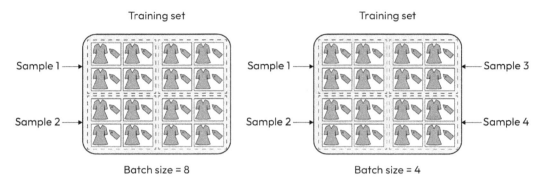

Figure 1.3 – Concept of batch size

With a batch size equal to eight, the dataset in the example is divided into two samples, each with eight dataset instances. On the other hand, with a lower batch size (in this case, four), the training set is divided into a higher number of samples (four samples).

The neural network receives input samples and outputs a set of results, each corresponding to an instance of the input sample. In the case of a model to treat the classification image problem of Fashion-MNIST, the neural network gets a set of images and outputs another set of labels, as you can see in *Figure 1.4*. Each one of these labels indicates the corresponding class of the input image:

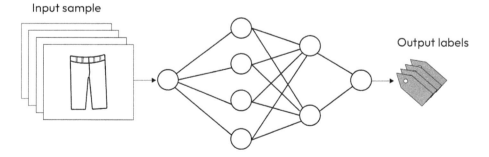

Figure 1.4 – Neural networks work on input samples

To extract the intrinsic knowledge present in the dataset, we need to submit the neural network to a training algorithm so it can learn the pattern present in the data. Let's jump to the next section to understand how this algorithm works.

The training algorithm

The training algorithm is an **iterative process** that takes each dataset sample and adjusts the neural network parameters according to the degree of error related to the difference between the correct result and the predicted one.

A single training iteration is called the **training step**. So, the number of training steps executed in the learning process equals the number of samples used to train the model. As we stated before, the batch size defines the number of samples, which also determines the number of training steps.

After executing all the training steps, we say the training algorithm has completed a **training epoch**. The developer must define the number of epochs before starting the model-building process. Usually, the developer determines the number of epochs by varying it and evaluating the accuracy of the resultant model.

A single training step executes the four phases sequentially, as illustrated in *Figure 1.5*:

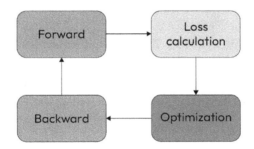

Figure 1.5 – The four phases of the training process

Let's go through each one of these steps to understand their role in the entire training process.

Forward

In the forward phase, the neural network receives the input data, performs calculations, and outputs a result. This output is also known as the value predicted by the neural network. In the case of Fashion-MNIST, the input data is the grayscale image and the predicted value is the class to which the item belongs.

Considering the tasks executed in the training step, the forward phase has a higher computational cost. This happens because it executes all the heavy computations involved in the neural network. Such computations, commonly known as operations, will be explained in the next section.

It is interesting to note that the forward phase is exactly the same as the inference process. When using the model in practice, we continuously execute the forward phase to infer a value or result.

Loss calculation

After the forward phase, the neural network will output a predicted value. Then, the training algorithm needs to compare the predicted value with the expected one to see how good the prediction made by the model is.

If the predicted value is close or equal to the real value, the model is performing as expected and the training process is going in the right direction. Otherwise, the training step needs to quantify the error achieved by the model to adjust the parameters proportionally to the error degree.

> **Important note**
> In the terminology of neural networks, this error is usually referred to as **loss** or **cost**. So, it is common to see names such as loss or cost function in the literature when addressing this topic.

There are different kinds of loss functions, each one suitable to treat a specific sort of problem. The **cross-entropy** (**CE**) loss function is used in multiclass image classification problems, where we need to classify an image within a group of classes. For example, this loss function can be used in the Fashion-MNIST problem. Suppose we have just two classes or categories. In that case, we face a binary class problem, so using the **binary cross-entropy** (**BCE**) function rather than the original cross-entropy loss function is recommended.

For regression problems, the loss function is completely different from the ones used in classification problems. We can use functions such as the **mean squared error** (**MSE**), which measures the squared difference between the original value and the predicted value by the neural network.

Optimization

After obtaining the loss, the training algorithm calculates the partial derivative of the loss function concerning the current parameters of the network. This operation results in the so-called **gradient**, which the training process uses to adjust network parameters.

Leaving the mathematical foundations aside, we can think of the gradient as the change we need to apply to network parameters to minimize the error or loss.

> **Important note**
> You can find more information about the math used in deep learning by reading the book *Hands-On Mathematics for Deep Learning*, published by Packt and written by Jay Dawani.

Similar to the loss function, we also have distinct implementations of optimizers. The **stochastic gradient descent (SGD)** and Adam are used the most.

Backward

To finish the training process, the algorithm updates the network parameters according to the gradient obtained in the optimization phase.

> **Important note**
>
> This section provides a theoretical explanation of the training algorithm. So, be aware that depending on the machine learning framework, the training process can have a set of phases that is different from the ones in the preceding list.

Essentially, these phases constitute the computational burden of the training process. Follow me to the next section to understand how this computational burden is impacted by different factors.

Understanding the computational burden of the model training phase

Now that we've brushed up on how the training process works, let's understand the computational cost required to train a model. By using the terms computational cost or burden, we mean the computing power needed to execute the training process. The higher the computational cost, the higher the time taken to train the model. In the same way, the higher the computational burden, the higher the computing resources required to train the model.

Essentially, we can say the computational burden to train a model is defined by a three-fold factor, as illustrated in *Figure 1.6*:

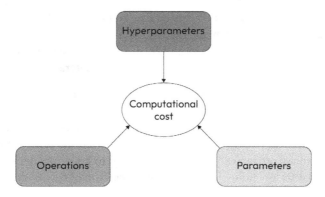

Figure 1.6 – Factors that influence the training computational burden

Each one of these factors contributes (to some degree) to the computational complexity imposed by the training process. Let's talk about each one of them.

Hyperparameters

Hyperparameters define two aspects of neural networks: the neural network configuration and how the training algorithm works.

Concerning neural network configuration, the hyperparameters determine the number and type of layers and the number of neurons in each layer. Simple networks have a few layers and neurons, whereas complex networks have thousands of neurons spread in hundreds of layers. The number of layers and neurons determines the number of parameters of the network, which directly impacts the computational burden. Due to the significant influence of the number of parameters in the computational cost of the training step, we will discuss this topic later in this chapter as a separate performance factor.

Regarding how the training algorithm executes the training process, hyperparameters control the number of epochs and steps and determine the optimizer and loss function used during the training phase, among other things. Some of these hyperparameters have a tiny influence on the computational cost of the training process. For example, if we change the optimizer from SGD to Adam, we will not face any relevant impact on the computational cost of the training process.

Other hyperparameters can definitely raise the training phase time, though. One of the most emblematic examples is the batch size. The higher the batch size, the fewer training steps are needed to train a model. So, with a few training steps, we can speed up the building process since the training phase will execute fewer steps per epoch. On the other hand, we can spend more time executing a single training step if we have big batch sizes. This happens because the forward phase executed on each training step should deal with a higher dimensional input data. In other words, we have a trade-off here.

For example, consider the case of a batch size equal to *32* for the Fashion-MNIST dataset. In this case, the input data dimension is *32 x 1 x 28 x 28*, where 32, 1, and 28 represent the batch size, the number of channels (colors, in this scenario), and the image size, respectively. Therefore, for this case, the input data comprises 25,088 numbers, which is the number of numbers the forward phase should compute. However, if we increase the batch size to *128*, the input data changes to 100,352 numbers, which can result in a longer time to execute a single forward phase iteration.

In addition, a bigger input sample requires a higher amount of memory to execute each training step. Depending on the hardware configuration, the amount of memory required to execute the training step can drastically reduce the performance of the entire training process or even make it impossible to execute in that hardware. Conversely, we can accelerate the training process by using hardware endowed with huge memory resources. This is why we need to know the details of the hardware resources we use and what factors influence the computational complexity of the training process.

We will dive into all of these issues throughout the book.

Operations

We already know each training step executes four training phases: forward, loss computation, optimization, and backward. In the forward phase, the neural network receives the input data and processes it according to the neural network's architecture. Besides other things, the architecture defines the network layers, where each layer has one or more operations that the network executes during the forward phase.

For example, a **fully connected neural network (FCNN)** usually executes general matrix-to-matrix multiplication operations, whereas **convolutional neural networks (CNNs)** execute special computer vision operations such as convolution, padding, and pooling. It turns out that the computational complexity of one operation is not the same as another. So, depending on the network architecture and the operations, we can get distinct performance behavior.

Nothing is better than an example, right? Let's define a class to instantiate a traditional CNN model that is able to deal with the Fashion-MNIST dataset.

> **Important note**
>
> The complete code shown in this section is available at https://github.com/
> PacktPublishing/Accelerate-Model-Training-with-PyTorch-2.X/
> blob/main/code/chapter01/cnn-fashion_mnist.ipynb.

This model receives an input sample of size *64 x 1 x 28 x 28*. This means the model receives 64 grayscale images (one channel) with a height and width equal to 28 pixels. As a result, the model outputs a tensor of dimension *64 x 10*, which represents the probability of the image belonging to each of the 10 categories of the Fashion-MNIST dataset.

The model has two convolutional and two fully connected layers. Each convolutional layer comprises one bidimensional convolution, the **rectified linear unit (ReLU)** activation function, and pooling. The first fully connected layer has 3,136 neurons connected to the 512 neurons of the fully connected second layer. The second layer is then connected to the 10 neurons of the output layer.

> **Important note**
>
> If you are unfamiliar with CNN models, it would be useful to watch the video *What is a convolutional neural network (CNN)?* from the Packt YouTube channel at https://youtu.
> be/K_BHmztRTpA.

By exporting this model to the ONNX format, we get the diagram illustrated in *Figure 1.7*:

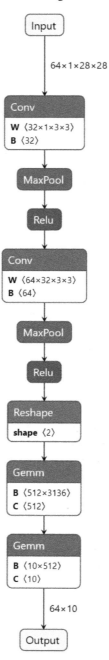

Figure 1.7 – Operations of a CNN model

> **Important note**
>
> The **Open Neural Network Exchange** (**ONNX**) is an open standard for machine learning interoperability. Besides other things, ONNX provides a standard format to export neural network models from many distinct frameworks and tools. We can use the ONNX file to inspect model details, import it into another framework, or execute the inference process.

By evaluating *Figure 1.7*, we can see five distinct operations:

- `Conv`: Bidimensional convolution
- `MaxPool`: Max pooling
- `Relu`: Activation function (ReLU)
- `Reshape`: Tensor dimensional transformation
- `Gemm`: General matrix multiplication

So, under the hood, the neural network executes these operations in the forward phase. From a computing perspective, this is the set of real operations that the machine runs during each training step. Therefore, we can rethink the training process of this model in terms of its operations and write it as a simpler algorithm:

```
for each epoch
    for each training step
        result = conv(input)
        result = maxpool(result)
        result = relu(result)
        result = conv(result)
        result = maxpool(result)
        result = relu(result)
        result = reshape(result)
        result = gemm(result)
        result = gemm(result)
    loss = calculate_loss(result)
    gradient = optimization(loss)
    backwards(gradient)
```

As you can see, the training process is just a set of operations executed one after another. Despite the functions or classes used to define the model, the machine is actually running this set of operations.

It turns out that each operation has a particular computational complexity, thus requiring distinct levels of computing power and resources to be executed satisfactorily. In this way, we can face different performance gains and bottlenecks for each one of those operations. Similarly, some operations can be more suitable to execute in a given hardware architecture, as we will see throughout the book.

To obtain the practical meaning of this topic, we can check the percentage of time these operations spent during the training step. So, let's use **PyTorch Profiler** to get the percentage of CPU usage for each operation. The following list resumes the CPU usage when running the forward phase of our CNN model with one input sample of the Fashion-MNIST dataset:

```
aten::mkldnn_convolution: 44.01%
aten::max_pool2d_with_indices: 30.01%
aten::addmm: 13.68%
aten::clamp_min: 6.96%
aten::convolution: 1.18%
aten::copy_: 0.70%
aten::relu: 0.59%
aten::_convolution: 0.49%
aten::empty: 0.35%
aten::_reshape_alias: 0.31%
aten::t: 0.31%
aten::conv2d: 0.24%
aten::as_strided_: 0.24%
aten::reshape: 0.21%
aten::linear: 0.21%
aten::max_pool2d: 0.17%
aten::expand: 0.14%
aten::transpose: 0.10%
aten::as_strided: 0.07%
aten::resolve_conj: 0.00%
```

> **Important note**
>
> ATen is a C++ library used by PyTorch to execute basic operations. You can find more information about this library at https://pytorch.org/cppdocs/#aten.

The results show the Conv operation (labeled here as aten::mkldnn_convolution) presented higher CPU usage (44%), followed by the MaxPool operation (aten:: max_pool2d_with_indices), with 30% CPU usage. On the other hand, the ReLU (aten::relu) and Reshape (aten::reshape) operations consumed less than 1% of the total CPU usage. Finally, the Gemm operation (aten::addmm) used around 14% of the CPU time.

From this simple profiling test, we can assert the operations involved in the forward phase; hence, in the training process, there are distinct levels of computational complexity. We can see the training process consumed much more CPU cycles when executing the Conv operation than the Gemm operation. Notice that our CNN model has two layers comprising both operations. Thus, in this example, both operations are executed the same number of times.

Based on this knowledge about the distinct computational burden of neural network operations, we can choose the best hardware architecture or software stack to reduce the execution time of the predominant operation of a given neural network. For example, suppose we need to train a CNN composed of dozens of convolutional layers. In that case, we will look for hardware resources endowed with special capabilities to execute Conv operations more efficiently. Even though the model has some fully connected layers, we already know that the Gemm operation can be less computationally intensive than Conv. This justifies prioritizing a hardware resource that is able to accelerate convolutional operations to train that model.

Parameters

Besides hyperparameters and operations, the neural network parameters are another factor that has a relevant influence on the computational cost of the training process. As we discussed earlier, the number and type of layers in the neural network configuration define the total number of parameters on the network.

Obviously, the higher the number of parameters, the higher the computational burden of the training process. These parameters comprise kernel values employed on convolutional operations, biases, and the weights of connections between neurons.

Our CNN model, with just 4 layers, has 1,630,090 parameters. We can easily count the total number of parameters in PyTorch by using this function:

```
def count_parameters(model):
    parameters = list(model.parameters())
    total_parms = sum(
        [np.prod(p.size()) for p in parameters if p.requires_grad])
    return total_parms
```

If we add an extra fully connected layer with 256 neurons to our CNN model and rerun this function, we will get 1,758,858 parameters in total, representing an increase of nearly 8%.

After training and testing this new CNN model, we got the same accuracy as before. Then, paying attention to the trade-off between network complexity and model accuracy is essential. On many occasions, increasing the number of layers and neurons will not necessarily result in better efficiency but will possibly increase the time of the training process.

Another aspect of parameters is the numeric precision used to represent these numbers in the model. We will dive into this topic in *Chapter 7, Adopting Mixed Precision*, but for now, keep in mind that the number of bytes used to represent parameters pays a relevant contribution to the time needed to train a model. So, not only does the number of parameters have an impact on training time but so does the numeric precision chosen to represent these numbers in the model.

The next section brings a couple of questions to help you retain what you have learned in this chapter.

Quiz time!

Let's review what we have learned in this chapter by answering eight questions. At first, try to answer these questions without consulting the material.

> **Important note**
> The answers to all these questions are available at `https://github.com/PacktPublishing/Accelerate-Model-Training-with-PyTorch-2.X/blob/main/quiz/chapter01-answers.md`.

Before starting the quiz, remember that it is not a test at all! This section aims to complement your learning process by revising and consolidating the content covered in this chapter.

Choose the correct option for the following questions:

1. Which phases comprise the training process?

    A. Forward, processing, optimization, and backward.

    B. Processing, pre-processing, and post-processing.

    C. Forward, loss calculation, optimization, and backward.

    D. Processing, loss calculation, optimization, and post-processing.

2. Which factors impact the computational burden of the training process?

    A. Loss function, optimizer, and parameters.

    B. Hyperparameters, parameters, and operations.

    C. Hyperparameters, loss function, and operations.

    D. Parameters, operations, and loss function.

3. After executing the training algorithm on all dataset samples, the training process has completed a training what?

    A. Evolution.

    B. Epoch.

    C. Step.

    D. Generation.

4. A dataset sample comprises a set of what?

    A. Dataset collections.

    B. Dataset steps.

    C. Dataset epochs.

    D. Dataset instances.

5. Which hyperparameter is more likely to increase the computational burden of the training process?

    A. Batch size.

    B. Optimizer.

    C. Number of epochs.

    D. Learning rate.

6. A training set has 2,500 instances. By defining a batch size equal to 1 and 50, the number of steps executed during the training process is, respectively, which of the following?

    A. 500 and 5.

    B. 2,500 and 1.

    C. 2,500 and 50.

    D. 500 and 50.

7. The profiling of a training process showed that the most time-consuming operation was `aten::mkldnn_convolution`. In this case, what is the heavier computing phase of the training process?

    A. Backward.

    B. Forward.

    C. Loss calculation.

    D. Optimization.

8. A model has two convolutional layers and two fully connected layers. If we add two more convolutional layers to the model, it will increase the number of what?

    A. Hyperparameters.

    B. Training steps.

    C. Parameters.

    D. Training samples.

Let's summarize what we've learned in this chapter.

Summary

We have reached the end of the first step of our training acceleration journey. You started this chapter by remembering how the training process works. In addition to refreshing concepts such as datasets and samples, you remembered the four phases of the training algorithm.

Next, you learned that hyperparameters, operations, and parameters are the three-fold factors influencing the training process's computational burden.

Now that you have remembered the training process and understood what contributes to its computational complexity, it's time to move on to the next topic.

Let's take our first steps to learn how to accelerate this heavy computational process!

2

Training Models Faster

In the last chapter, we learned the factors that contribute to increasing the computational burden of the training process. Those factors have a direct influence on the complexity of the training phase and, hence, on the execution time.

Now, it is time to learn how to accelerate this process. In general, we can improve performance by changing something in the software stack or increasing the number of computing resources.

In this chapter, we will start to understand both of these options. Next, we will learn what can be modified in the application and environment layers.

Here is what you will learn as part of this chapter:

- Understanding the approaches to accelerate the training process
- Knowing the layers of the software stack used to train a model
- Learning the difference between vertical and horizontal scaling
- Understanding what can be changed in the application layer to accelerate the training process.
- Understanding what can be changed in the environment layer to improve the performance of the training phase

Technical requirements

You can find the complete code of the examples mentioned in this chapter in the book's GitHub repository at https://github.com/PacktPublishing/Accelerate-Model-Training-with-PyTorch-2.X/blob/main.

You can access your favorite environments to execute this notebook, such as Google Colab or Kaggle.

What options do we have?

Once we have decided to accelerate the training process of a model, we can take two directions, as illustrated in *Figure 2.1*:

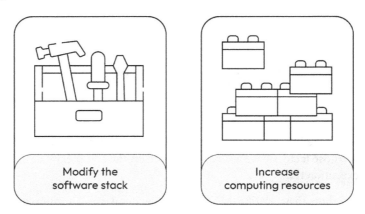

Figure 2.1 – Approaches to accelerating the training phase

In the first option (**Modify the software stack**), we go through each layer of the software stack used to train a model to seek opportunities to improve the training process. In simpler words, we can change the application code, install and use a specialized library, or enable a special capability regarding the operating system or container environment.

This first approach relies on having profound knowledge of performance tuning techniques. In addition, it demands a high sense of investigation to identify bottlenecks and apply the most suitable solution to overcome them. Thus, this approach is about harnessing the most hardware and software resources by extracting the maximum performance of the computing system.

Nevertheless, remark that depending on the environment we are running the training process in, we may not have the required privileges to change the lower layers of the software stack. For example, suppose we are running the training process in a notebook provided by a third-party environment such as **Kaggle** or **Google Colab**. In this case, we cannot change operating system parameters or modify the container image because this environment is controlled and restricted. We can still change the application code, but it may not be enough to accelerate the training process.

When changing things in the software stack is impossible or does not provide the expected performance gain, we can go towards the second option (**Increase computing resources**) to train the model. So, we can increase the number of processors and the amount of main memory, use accelerator devices, or spread the training process across multiple machines. Naturally, we may need to spend money to bring this option to life.

Notice that this approach is easier in the cloud than for on-premises infrastructures. When using the cloud, we can easily contract a machine endowed with accelerator devices or add more machines to our setup. We can get these resources ready to use with a few clicks. On the other hand, we may face some constraints when adding new computing resources to the on-premises infrastructure, such as physical space and energy capacity restrictions. It is not impossible, though; it might only be more challenging to do.

Furthermore, we also have another scenario in which our infrastructure, either on the cloud or on-premises, already has those computing resources. In this case, we just need to start using them to accelerate the training process.

So, if we have money to buy or contract these resources, or if they are already available to us in our environment, the problem is solved, right? Not necessarily. Unfortunately, there is no guarantee that using additional resources in the training process will automatically improve performance. As we will discuss in this book, the performance bottleneck is not always overcome by adding more computing resources without rethinking the whole process, adjusting the code, and so on.

This last assertion gives us a valuable lesson: we must see these two approaches as a cycle, not as two isolated options. This means we must go back and forth on both methods as often as necessary to reach the desired improvement, as shown in *Figure 2.2*:

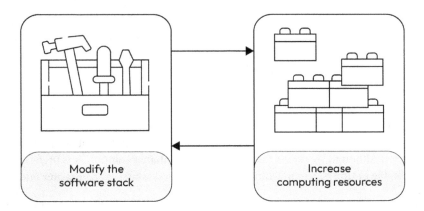

Figure 2.2 – Continuous improvement cycle

Let's see more details about both approaches in the following sections.

Modifying the software stack

The software stack used to train a model can vary depending on the environment we use to execute this process. For the sake of simplicity, we will consider in this book a software stack seen from the point of view of data scientists, i.e., as users of a computing service or environment.

In general, we can say the software stack looks like the layers shown in *Figure 2.3*:

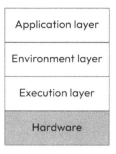

Figure 2.3 – Software stack used to train a model

From the top to the bottom, we have the following layers:

1. **Application**: The model-building program occupies this layer. This program can be written in any programming language capable of building neural networks, such as R and Julia, but Python is the language primarily used for this purpose.

2. **Environment**: The machine learning framework used to build the application, libraries, and tools used to support this framework lie in this layer. Some examples of machine learning frameworks are **PyTorch**, **TensorFlow**, **Keras**, and **MxNet**. Concerning the set of libraries, we can cite **Nvidia Collective Communication Library** (**NCCL**), for efficient communication among GPUs, and **jemalloc**, for optimized memory allocation.

3. **Execution**: This layer is responsible for supporting the execution of the environment and application layers. Therefore, a container solution or bare metal operating system belongs to this layer. Although the components of the upper layers can be executed directly on the operating system, nowadays, it is common to use a container to wrap up the entire application and its environment. Although Docker is the most famous container solution, it is preferable to adopt a more suitable option to run machine learning workloads such as **Apptainer** and **Enroot**.

At the bottom of the software stack, there is a box representing all the hardware resources needed to execute the upper software layers.

To obtain a practical understanding of this software stack representation, let's see a couple of examples, as illustrated in *Figure 2.4*:

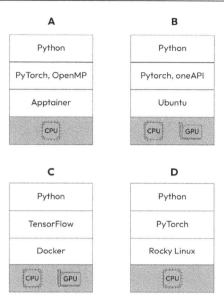

Figure 2.4 – Examples of software stacks

All scenarios described in *Figure 2.4* use an application written in Python. As stated before, the application can be a program coded in C++ or a script written in R. It does not matter. The important note to keep in mind is that the application layer represents the location of our code. In examples **A**, **B**, and **D**, we have scenarios using PyTorch as the machine learning framework. Cases **A** and **B** have the support of additional libraries, namely **OpenMP** and **Intel One API**. This means PyTorch is relying on these libraries to empower tasks and operations.

Finally, the execution layer of scenarios **B** and **C** uses a container solution to execute the upper layers, whereas the upper layers of examples **A** and **D** run directly on the operating system. Furthermore, notice that the hardware resources in scenarios **B** and **C** are endowed with GPU accelerators, whereas the others have only a CPU.

> **Important note**
> Remark that we are abstracting the type of infrastructure used to run the software stack since it is irrelevant to our discussion at this moment. Therefore, you can consider the software stack hosted in a cloud or on-premises infrastructure.

Except for the case where we are using an environment provided by our own resources, we probably will not have the administrative rights to modify or add configurations in the execution layer. In most cases, we use the computing environments provisioned by our companies. Thus, we do not possess the privilege to alter anything in the container or operating system layers. Usually, we address these modifications to the IT infrastructure team.

For this reason, we will focus on the application and environment layers, where we have the power to change and perform additional configurations.

The second part of this book focuses on teaching you how to change the software stack in such a way that we can accelerate the training process with the available resources.

> **Important note**
>
> There are exciting configurations we can make in the execution layer to improve performance. However, they are out of the scope of this book. As data scientists are the primary audience of this material, we focus on the layers that these professionals have the necessary access to modify and customize by themselves.

Modifying the software stack to accelerate the training process has a limit. We get stuck on performance improvement, no matter how deep and advanced the techniques we use are. When we reach that limit, the one way to speed up the training phase is by using additional computing resources, as explained in the next section.

Increasing computing resources

There are two approaches to increasing computing resources in an existing environment: vertical and horizontal scaling. In vertical scaling, we augment the computing resources of a single machine, whereas, in horizontal scaling, we add more machines to the pool of equipment used to train the model.

In practical terms, **vertical scaling** allows for equipping the machine with an accelerator device to increase main memory, add more processor cores, and so on, as exemplified in *Figure 2.5*. After doing this scaling, we obtain an empowered machine with higher resources:

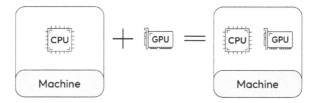

Figure 2.5 – Example of vertical scaling

Horizontal scaling is related to the increase in the number of machines used by our application. If we originally used one machine to execute the training process, we can apply horizontal scaling and use two machines to work together to train the model, as shown in the example of *Figure 2.6*:

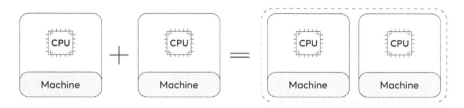

Figure 2.6 – Example of horizontal scaling

Regardless of the type of scaling, we need to know how to harness these additional resources to improve performance. Depending on the kind of resources added to our setup, we need to adjust the code in many different parts. In other situations, the machine learning framework can automatically deal with the increase in resources without requiring any additional modification.

As we learned in this section, the first step to accelerate the training process relies on modifying the application layer. Follow me to the next section to know how to do it.

Modifying the application layer

The application layer is the starting point of the performance improvement journey. As we have complete control of the application code, we can change it without depending on anyone else. Thus, there is no better way to start the performance optimization process than working independently.

What can we change in the application layer?

You may wonder how we can modify the code to improve performance. Well, we can reduce model complexity, increase the batch size to optimize memory usage, compile the model to fuse operations and disable profiling functions to eliminate extra overhead in the training process.

Regardless of the changes applied to the application layer, we cannot sacrifice model accuracy in favor of performance improvement since this does not make sense. As the primary goal of a neural network is to solve problems, it would be meaningless to accelerate the building process of a useless model. Then, we must pay attention to model quality when modifying the code to reduce the training phase time.

In *Figure 2.7*, we can see the sort of changes we can make in the application layer to speed up the training phase:

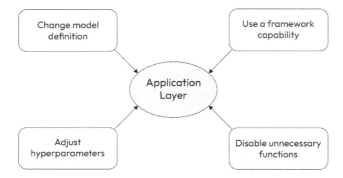

Figure 2.7 – Changes in the application layer to accelerate the training process

Let's look at each of the changes:

- **Change model definition**: Modify the neural network architecture to reduce the number of layers, weights, and operations executed on each layer

- **Adjust hyperparameters**: Change hyperparameters such as batch size, the number of epochs, and the optimizer

- **Use a framework capability**: Take advantage of a framework capability such as kernel fusion, automatic mixed precision, and model compiling

- **Disable unnecessary functions**: Get rid of undue burdens such as computing the gradient on the validation phase

> **Important note**
> Some framework capabilities rely on making changes in the environment layer, such as installing an additional tool or library or even upgrading the framework version.

Naturally, these categories do not cover all possibilities for performance improvement in the application layer; their purpose is to give you a clear mental model of what we can effectively do to the code to accelerate the training phase.

Getting hands-on

Let's see a practical example of performance improvement by changing only the application code. Our guinea pig is the CNN model introduced in the previous chapter, which was used to classify the images of the Fashion-MNIST dataset.

> **Important note**
>
> Details about the computing environment used in this experiment are irrelevant at this time. What truly matters is the speedup achieved with these modifications, considering the same environment and conditions.

This model has two convolutional layers and two fully connected layers, resulting in **1,630,090** weights. With the number of epochs equal to 10 and the batch size equal to 64, the training phase took 148 seconds to complete. The trained model achieved 83.99% accuracy when tested against 10,000 images from the test dataset, as you can see here:

```
Epoch [1/10], Loss: 0.9136
Epoch [2/10], Loss: 0.6925
Epoch [3/10], Loss: 0.7313
Epoch [4/10], Loss: 0.6681
Epoch [5/10], Loss: 0.3191
Epoch [6/10], Loss: 0.5790
Epoch [7/10], Loss: 0.4824
Epoch [8/10], Loss: 0.6229
Epoch [9/10], Loss: 0.7279
Epoch [10/10], Loss: 0.3292
Training time: 148 seconds
Accuracy of the network on the 10000 test images: 83.99 %
```

> **Important note**
>
> The complete code shown in this section is available at https://github.com/ PacktPublishing/Accelerate-Model-Training-with-PyTorch-2.X/ blob/main/code/chapter02/baseline.ipynb.

By making only one simple modification to the code, we can reduce the training time of this model by 15% while keeping the same accuracy achieved with the baseline code. The improved code took 125 seconds to complete, and the trained model reached an accuracy equal to 83.76%:

```
Epoch [1/10], Loss: 1.0960
Epoch [2/10], Loss: 0.6656
Epoch [3/10], Loss: 0.6444
Epoch [4/10], Loss: 0.6463
Epoch [5/10], Loss: 0.4772
Epoch [6/10], Loss: 0.5548
Epoch [7/10], Loss: 0.4800
Epoch [8/10], Loss: 0.4190
Epoch [9/10], Loss: 0.4885
```

```
Epoch [10/10], Loss: 0.4708
Training time: 125 seconds
Accuracy of the network on the 10000 test images: 83.76 %
```

> **Important note**
>
> The complete code shown in this section is available at `https://github.com/PacktPublishing/Accelerate-Model-Training-with-PyTorch-2.X/blob/main/code/chapter02/application_layer-bias.ipynb`.

We improved performance by disabling the bias parameter on the two convolutional and two fully connected layers. The following piece of code shows the use of the `bias` parameter to disable the bias weight on the function's `Conv2d` and `Linear` layers:

```
def __init__(self, num_classes=10):
    super(CNN, self).__init__()
    self.layer1 = nn.Sequential(
        nn.Conv2d
            (1, 32, kernel_size=3, stride=1,padding=1, bias=False),
        nn.ReLU(),
        nn.MaxPool2d(kernel_size = 2, stride = 2))

    self.layer2 = nn.Sequential(
        nn.Conv2d
            (32, 64, kernel_size=3,stride=1,padding=1, bias=False),
        nn.ReLU(),
        nn.MaxPool2d(kernel_size = 2, stride = 2))

    self.fc1 = nn.Linear(64*7*7, 512, bias=False)
    self.fc2 = nn.Linear(512, num_classes, bias=False)
```

This modification reduced the number of weights from 1,630,090 to 1,629,472, representing a decrease of only 0.04% in the total number of neural network weights. As we can see, this change in the number of weights did not affect the model's accuracy since it achieved practically the same efficiency as before. Therefore, we trained the model 15% faster with almost no additional effort.

What if we change the batch size?

If we double the batch size from 64 to 128, we achieve an even better performance gain than disabling the bias:

```
Epoch [1/10], Loss: 1.1859
Epoch [2/10], Loss: 0.7575
Epoch [3/10], Loss: 0.6956
```

```
Epoch [4/10], Loss: 0.6296
Epoch [5/10], Loss: 0.6997
Epoch [6/10], Loss: 0.5369
Epoch [7/10], Loss: 0.5247
Epoch [8/10], Loss: 0.5866
Epoch [9/10], Loss: 0.4931
Epoch [10/10], Loss: 0.4058
Training time: 96 seconds
Accuracy of the network on the 10000 test images: 82.14 %
```

> **Important note**
>
> The complete code shown in this section is available at `https://github.com/PacktPublishing/Accelerate-Model-Training-with-PyTorch-2.X/blob/main/code/chapter02/application_layer-batchsize.ipynb`.

We trained the model 54% faster by doubling the batch size. As we learned in *Chapter 1, Deconstructing the Training Process*, the batch size dictates the number of steps in the training phase. Because we increased the batch size from 64 to 128, we obtained a fewer number of steps per epoch, i.e., the number of steps passed from 938 to 469. As a consequence, the learning algorithm executes half of the phases needed to complete an epoch.

However, such modification came with a price: the accuracy was reduced from 83.99% to 82.14%. This happens because the learning algorithm executes the optimization phase per each training step. Since the number of steps reduced and the number of epochs remained the same, the learning algorithm executed a fewer number of optimization phases, which consequently decreases its opportunity to reduce the training cost.

Just out of curiosity, let's see what happens when changing the batch size to `256`:

```
Epoch [1/10], Loss: 1.5919
Epoch [2/10], Loss: 0.9232
Epoch [3/10], Loss: 0.8151
Epoch [4/10], Loss: 0.6488
Epoch [5/10], Loss: 0.7208
Epoch [6/10], Loss: 0.5085
Epoch [7/10], Loss: 0.5984
Epoch [8/10], Loss: 0.5603
Epoch [9/10], Loss: 0.6575
Epoch [10/10], Loss: 0.4694
Training time: 76 seconds
Accuracy of the network on the 10000 test images: 80.01 %
```

The training time was reduced even more, though not as significantly as when we changed from 64 to 128. On the other hand, the model efficiency fell to 80%. We can also observe an increase in the loss per epoch when compared to the previous test.

In short, we have to find the balance between training speedup and model efficiency when adjusting the batch size. The ideal batch size value depends on the model architecture, dataset characteristics, and the hardware resource used to train the model. Thus, the best way to define it is by doing some experiments before starting the training process for real.

These simple examples showed is possible to accelerate the training process by making direct modifications to the code. In the next section, we will see what kind of changes we can make in the environment layer to speed up model training.

Modifying the environment layer

The environment layer comprises the machine learning framework and all the software needed to support its execution, such as libraries, compilers, and auxiliary tools.

What can we change in the environment layer?

As we discussed before, we may not have the necessary permission to change anything in the environment layer. This restriction depends on the type of environment we use to train the model. In third-party environments, such as notebook's online services, we do not have the flexibility to make advanced configurations, such as downloading, compiling, and installing a specialized library. We can upgrade a package or install a new library, but nothing beyond that.

To overcome this restriction, we commonly use **containers**. Containers allow us to configure anything we need to run our application without requiring the support or permission of everyone else. Obviously, we are talking about the environment layer and not about the execution layer. As we discussed previously, making changes to the execution layer requires administrative privileges, which would be out of our hands in most environments we usually use.

> **Important note**
> The complete code shown in this section is available at `https://github.com/PacktPublishing/Accelerate-Model-Training-with-PyTorch-2.X/blob/main/code/chapter02/environment_layer.ipynb`.

In the case of the environment layer, we can modify these sorts of things:

- **Install and use a specialized library**: The machine learning framework comes with everything we need to train a model. However, we can speed up the training process by using libraries specialized in tasks such as memory allocation, math operations, and collective communication.

- **Control libraries' behavior through environment variables**: The default behavior of libraries cannot be the best one for a given scenario or specific setup. In that case, we can change it directly through environment variables from the application code.

- **Upgrade framework and libraries to new versions**: This may sound silly, but upgrading the machine learning framework and libraries to new versions can raise the performance of the training process much more than we think.

We are going to learn many of these things throughout this book. For now, let's jump to the next section to see performance improvement in practice.

Getting hands-on

As we did in the last section, we will use the baseline code here to assess the performance gain from modifying the environment layer. Remember that the training process of our baseline code took 148 seconds to run. The environment layer used for that execution is composed of PyTorch 2.0 (2.0.0+cpu) as the machine learning framework.

After making two modifications to the environment layer, we got a performance improvement near 40%, with the model's accuracy being practically the same as before, as you can see:

```
Epoch [1/10], Loss: 0.6036
Epoch [2/10], Loss: 0.3941
Epoch [3/10], Loss: 0.4808
Epoch [4/10], Loss: 0.5834
Epoch [5/10], Loss: 0.6347
Epoch [6/10], Loss: 0.3218
Epoch [7/10], Loss: 0.4646
Epoch [8/10], Loss: 0.4960
Epoch [9/10], Loss: 0.3683
Epoch [10/10], Loss: 0.6173
Training time: 106 seconds
Accuracy of the network on the 10000 test images: 83.25 %
```

We made only one change to accelerate the training process of the baseline model by almost 40%: the installation and configuration of Intel OpenMP version 2023.1.0. We have configured the behavior of this library by setting three environment variables:

```
import os
os.environ['OMP_NUM_THREADS'] = "16"
os.environ['KMP_AFFINITY'] = "granularity=fine,compact,1,0"
os.environ['KMP_BLOCKTIME'] = "0"
```

In short, these parameters control the way Intel Open launches and orchestrates the threads, besides determining the number of threads created by the library. We should configure these parameters, taking into account the characteristics of the training burden and hardware resources. Notice that setting up these parameters in the code is part of modifying the environment layer and not the application layer. Even though we are changing the code, those modifications are related to environment control rather than model definition.

> **Important note**
> Do not worry about how to install and enable the Intel OpenMP library and what each of the environment variables used in this test means. We will cover this topic in detail in *Chapter 4, Using Specialized Libraries*.

Although the PyTorch package installed from PIP comes with the GNU OpenMP library by default, the Intel version tends to provide better results in machines endowed with Intel CPUs. As the hardware machine used in this test possesses an Intel CPU, it is recommended to use the Intel version of OpenMP instead of the implementation provided by the GNU project.

We can see that changing a few things in the environment layer can result in a relevant performance gain without consuming much time or effort to implement.

The next section provides some questions to help you retain what you have learned in this chapter.

Quiz time!

Let's review what we have learned in this chapter by answering a few questions. At first, try to answer these questions without consulting the material.

> **Important note**
> The answers to all these questions are available at `https://github.com/PacktPublishing/Accelerate-Model-Training-with-PyTorch-2.X/blob/main/quiz/chapter02-answers.md`.

Before starting the quiz, remember that it is not a test at all! This section aims to complement your learning process by revising and consolidating the content covered in this chapter.

Choose the correct answers for the following questions:

1. After running the training process using two GPUs in a single machine, we decided to add two extra GPUs to accelerate the training process. In this case, we tried to improve the performance of the training process by applying which of the following?

    A. Horizontal scaling.

    B. Vertical scaling.

C. Transversal scaling.

D. Distributed scaling.

2. The training process of a simple model is taking a long time to finish. After adjusting the batch size and cutting of one of the convolutional layers, we could train the model faster while achieving the same accuracy. In this case, we improve the performance of the training process by changing which of the following layers of the software stack?

A. Application layer.

B. Hardware layer.

C. Environment layer.

D. Execution layer.

3. Which of the following changes is applied to the environment layer?

A. Modify the hyperparameters.

B. Adopt another network architecture.

C. Update the framework's version.

D. Set a parameter in the operating system.

4. Which one of the following components lies in the execution layer?

A. OpenMP.

B. PyTorch.

C. Apptainer.

D. NCCL.

5. As users of a given environment, we usually do not modify anything at the execution layer. What is the reason for that?

A. We usually do not have administrative rights to change anything at the execution layer.

B. There is no change at the execution layer that could accelerate the training process.

C. The execution and application layers are almost the same thing. So, there is no difference between changing one or another layer.

D. As we usually execute the training process on containers, there is no change on the execution layer that could improve the training process.

6. We have accelerated the training process of a given model by using two additional machines and applying a given capability provided by the machine learning framework. In this case, which of the following actions have we taken to improve the training process?

    A. We have performed horizontal and vertical scaling.

    B. We have performed horizontal scaling and increased the number of resources.

    C. We have performed horizontal scaling and applied changes to the environment layer.

    D. We have performed horizontal scaling and applied changes to the execution layer.

7. Controlling the behavior of a library through environment variables is a change that is applied in which of the following layers?

    A. Application layer.

    B. Environment layer.

    C. Execution layer.

    D. Hardware layer.

8. Increasing the batch size can improve the performance of the training process. However, it can also present which of the following side effects?

    A. Reduce the number of samples.

    B. Reduce the number of operations.

    C. Reduce the number of training steps.

    D. Reduce model accuracy.

Let's summarize what we've covered in this chapter.

Summary

We reached the end of the introductory part of the book. We started this chapter by learning the approaches we can take to reduce the training time. Next, we learned what kind of modifications we can perform in the application and environment layers to accelerate the training process.

We have experienced, in practice, how changing a few things in the code or environment can result in impressive performance improvements.

You are ready to move on in the performance journey! In the next chapter, you will learn how to apply one of the most exciting capabilities provided by PyTorch 2.0: model compilation.

Part 2: Going Faster

In this part, you will learn about the main techniques and approaches that can be used in PyTorch to accelerate the training process of deep learning models. First, you will learn how to compile a model by using the Compile API. After that, you will learn how to use and configure specialized libraries to optimize the training process on CPUs. Then, you will learn how to build an efficient data pipeline to keep the GPU busy all the time. Also, you will learn how to simplify the model by applying pruning and compression techniques. Finally, you will learn how to adopt automatic mixed precision to reduce computing time and memory consumption.

This part has the following chapters:

- *Chapter 3, Compiling the Model*
- *Chapter 4, Using Specialized Libraries*
- *Chapter 5, Building an Efficient Data Pipeline*
- *Chapter 6, Simplifying the Model*
- *Chapter 7, Adopting Mixed Precision*

3
Compiling the Model

Paraphrasing one of the famous presenters: "It's time!" After completing our initial steps toward performance improvement, it is time to learn a new capability of PyTorch 2.0 to accelerate the training and inference of deep learning models.

We are talking about the Compile API, which was presented in PyTorch 2.0 as one of the most exciting capabilities of this new version. In this chapter, we will learn how to use this API to build a faster model to optimize the execution of its training phase.

Here is what you will learn as part of this chapter:

- The benefits of graph mode over eager mode
- How to use the API to compile a model
- The components, workflow, and backends used by the API

Technical requirements

You can find the complete code for the examples mentioned in this chapter in this book's GitHub repository at `https://github.com/PacktPublishing/Accelerate-Model-Training-with-PyTorch-2.X/blob/main`.

You can access your favorite environment to execute this notebook, such as Google Collab or Kaggle.

What do you mean by compiling?

As a programmer, you will immediately assign the term "compiling" to the process of building a program or application from the source code. Although the complete building process comprises additional phases, such as generating assembly code and linking it to libraries and other objects, it is reasonable to think that way. However, at first glance, it may be a bit confusing to think about the compiling process in the context of this book since we are talking about Python. After all, Python is not a compiled language; it is an interpreted language, and thus, no compiling is involved.

> **Note**
>
> It is important to clarify that Python uses compiled functions for performance purposes, though it is primarily an interpreted language.

That said, what is the meaning of compiling a model? Before answering this question, we must understand the two execution modes of machine learning frameworks. Follow me to the next section.

Execution modes

Essentially, machine learning frameworks have two distinct execution modes. In **eager mode**, each operation is executed as it appears in the code, which is exactly what we expect to see in an interpreted language. The interpreter – Python, in this case – executes the operation as soon as it comes to light. So, there is no evaluation of what comes next when the operation is executed:

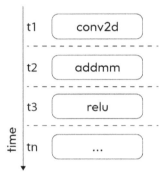

Figure 3.1 – Eager execution mode

As shown in *Figure 3.1*, the interpreter executes the three operations one after another in instants of t1, t2, and t3. The term "eager" stands for doing things instantly without taking a breath to evaluate the whole scenario before making the next step.

Besides eager mode, there is another approach named **graph mode**, which is similar to the traditional compiling process. Graph mode evaluates the complete set of operations to seek optimization opportunities. To perform this process, the program must evaluate the task as a whole, as shown in *Figure 3.2*:

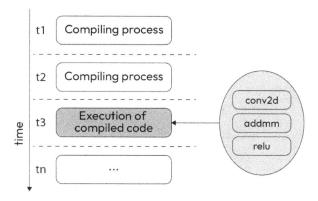

Figure 3.2 – Graph execution mode

Figure 3.2 shows the program uses t1 and t2 to execute the compiling process instead of eagerly running operations, as done before. The set of operations is executed only at t3 when the compiled code is executed.

The term "graph" refers to the directed graph created by this execution mode to represent operations and operands of a task. As this graph represents the processing flow of the task, the execution mode evaluates this representation to find ways to fuse, condense, and optimize operations.

For example, consider the case illustrated in *Figure 3.3*, which represents a task comprised of three operations. Op1 and Op2 receive operands I1 and I2, respectively. The results of these computations are used as input to Op3. Here, Op3 takes these two operands and, together with operand I3, outputs a result of O1:

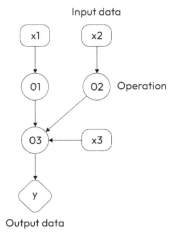

Figure 3.3 – Example of operations represented in a directed graph

After evaluating this graph, the program could decide to fuse all three operations in a single compiled code. As shown in *Figure 3.4*, this piece of code receives all three operands and outputs a value of O1:

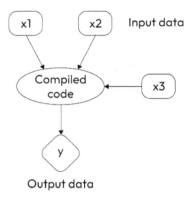

Figure 3.4 – Example of compiled operations

Besides fusing and reducing operations, the compiled model – the resultant of graph mode – can be created specifically for some hardware architecture to harness all resources and capabilities provided by that device. This is one of the reasons graph mode can perform better than eager mode.

As with everything in life, each mode has advantages and drawbacks. In short, eager mode is simpler to understand and hack, besides not having any delay in starting to run operations. On the other hand, graph mode is much faster to execute, though it is more complex to comprehend and requires extra initial time to create the compiled piece of code.

Model compiling

Now that you've been introduced to eager and graph modes, we can go back to the question that was posed at the beginning of this section: what is the meaning of compiling a model?

Compiling a model means *changing the execution mode of forward and backward phases from eager to graph mode*. In doing this, the machine learning framework evaluates all operations and operands involved in these phases in advance to compile them into a single piece of code. Therefore, notice that when we use the term "compiling a model," we are referring to compiling the processing flow that's executed in the forward and backward phases.

But why would we want to do that? We compile a model to accelerate its training time since the compiled code tends to **run faster** than the code that's executed in eager mode. As we will see in the next few sections, performance improvement depends on diverse factors, such as the computational capability of the GPU used to train the model.

Remark, however, that performance improvement is not guaranteed for all hardware platforms and models. On many occasions, the performance of graph mode can be the same as eager mode or even worse because of the extra time needed to compile the model. Even so, we should always consider compiling the model to verify the resultant performance improvement, especially when using novel GPU devices.

At this point, you may be wondering what execution mode is supported by PyTorch. The default execution mode of PyTorch is eager mode since it is "easier to use and more suitable for machine learning researchers," as stated on PyTorch's website. Nevertheless, PyTorch also supports graph mode! After version 2.0, PyTorch natively supports graph execution mode through the **Compile API**.

Before this new version, we needed to use third-party tools and libraries to enable graph mode on PyTorch. However, with the launch of the Compile API, we can now easily compile a model. Let's learn how to use this API to accelerate the training phase of our models.

Using the Compile API

We will start learning the basic usage of the Compile API by applying it to our well-known CNN model and Fashion-MNIST dataset. After that, we will accelerate a heavier model that's used to classify images from the CIFAR-10 dataset.

Basic usage

Instead of describing the API's components and explaining a bunch of optional parameters, let's dive into a simple example to show the basic usage of this capability. The following piece of code uses the Compile API to compile the CNN model presented in previous chapters:

```
model = CNN()
graph_model = torch.compile(model)
```

> **Note**
>
> The complete code shown in this section is available at `https://github.com/PacktPublishing/Accelerate-Model-Training-with-PyTorch-2.X/blob/main/code/chapter03/cnn-graph_mode.ipynb`.

To compile a model, we need to call a function named `compile`, passing the model as a parameter. Nothing else is necessary for the basic usage of this API. `compile` returns an object that will be compiled the first time it is called. The rest of the code remains exactly as before.

We can set the following environment variable to see whether the compiling process occurs:

```
import os
os.environ['TORCH_COMPILE_DEBUG'] = "1"
```

If so, we will see a lot of messages, as shown here:

```
[INFO] Step 1: torchdynamo start tracing forward
[DEBUG] TRACE LOAD_FAST self []
[DEBUG] TRACE LOAD_ATTR layer1 [NNModuleVariable()]
[DEBUG] TRACE LOAD_FAST x [NNModuleVariable()]
[DEBUG] TRACE CALL_FUNCTION 1 [NNModuleVariable(), TensorVariable()]
```

Another way to certify that we compiled the model successfully is by profiling the forward phase by using the PyTorch Profiler API:

```
from torch.profiler import profile, ProfilerActivity

activities = [ProfilerActivity.CPU]
prof = profile(activities=activities)
input_sample, _ = next(iter(train_loader))

prof.start()
model(input_sample)
prof.stop()

print(prof.key_averages().table(sort_by="self_cpu_time_total",
                                row_limit=10))
```

If the model was compiled successfully, the profiling would show a task labeled as CompiledFunction, as shown in the first line of the following output:

```
CompiledFunction: 55.50%
aten::mkldnn_convolution: 30.36%
aten::addmm: 8.25%
aten::convolution: 1.06%
aten::as_strided: 0.63%
aten::empty_strided: 0.59%
aten::empty: 0.43%
aten::expand: 0.27%
aten::resolve_conj: 0.20%
detach: 0.20%
aten::detach: 0.16%
```

The preceding output shows that CompiledFunction and aten::mkldnn_convolution took almost 86% of the time required to execute the forward phase. If we profile the model when it's executed in eager mode, we can easily identify which operations were fused and transformed into CompiledFunction:

```
aten::mkldnn_convolution: 38.87%
aten::max_pool2d_with_indices: 27.31%
```

```
aten::addmm: 17.89%
aten::clamp_min: 6.63%
aten::convolution: 1.88%
aten::relu: 0.87%
aten::conv2d: 0.83%
aten::reshape: 0.57%
aten::empty: 0.52%
aten::max_pool2d: 0.52%
aten::linear: 0.44%
aten::t: 0.44%
aten::transpose: 0.31%
aten::expand: 0.26%
aten::as_strided: 0.13%
aten::resolve_conj: 0.00%
```

By evaluating the profiling output of eager and graph modes, we can see that the compiling process fused nine operations into the CompiledFunction operation, as illustrated in *Figure 3.5*. As shown in this example, there are situations where the compiling process cannot compile all operations involved in the forward phase. This happens due to many reasons, such as data dependence:

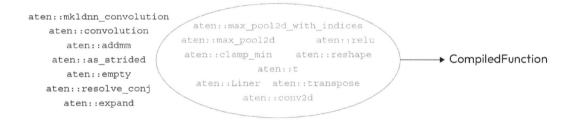

Figure 3.5 – A set of operations incorporated in a compiled function

You may be wondering about performance improvement. After all, this is what we came for! Do you remember what we discussed at the beginning of this chapter about not achieving performance improvement in all situations? Well, this is one of those cases.

Figure 3.6 shows the execution time of each training epoch of the CNN model running in both eager and graph modes. As we can see, the execution time of all epochs is higher in graph mode than in eager mode. Furthermore, the first epoch of graph mode is significantly slower than the others because the compiling process is executed at that moment:

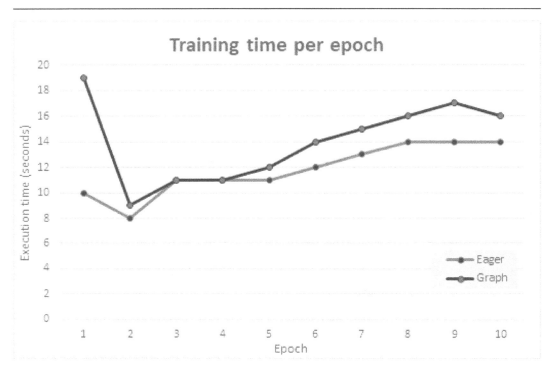

Figure 3.6 – Execution time of each training epoch of the CNN model in eager and graph modes

The overall time of model training executed on eager and graph modes was equal to 118 and 140 seconds, respectively. Thus, the compiled model was 18% slower than the default execution mode.

Frustrating, right? Yes, it is. However, remember that our CNN is just a toy model, thus there is not much space to truly improve performance. In addition, these experiments were executed in a non-GPU environment, though the compiling process tends to yield better results on GPU devices.

That said, let's go to the next section to see a significant performance improvement through the Compile API.

Give me a real fight – training a heavier model!

To see all the power of this capability in action, we will apply it to a more complex case. Our guinea pig for this is the **DenseNet121** model, which will be trained to classify images from the CIFAR-10 dataset. We can easily load and use both the model and the dataset from the torchvision module.

> **Note**
>
> The complete code shown in this section is available at https://github.com/PacktPublishing/Accelerate-Model-Training-with-PyTorch-2.X/blob/main/code/chapter03/densenet121_cifar10.ipynb

CIFAR-10 is a classic image classification dataset comprising 60,000 32x32 colored images. These images belong to 10 distinct categories, which explains the suffix "10" in the dataset's name.

Although each dataset image is 32x32 in size, it is a good practice to resize them to achieve better results on model training. Thus, we resized each image to 224x224 but kept the original three channels to represent the RGB color codification.

We conducted this experiment with the DenseNet121 model by using the following hyperparameters:

- **Batch size**: 64
- **Epochs**: 50
- **Learning rate**: 0.0001
- **Weight decay**: 0.005
- **Criterion**: Cross entropy
- **Optimizer**: Adam

Unlike previous experiments with the CNN model, this test was executed on an environment with the novel Nvidia A100 GPU. This GPU has a compute capability equal to 8.0, satisfying the requirement informed by PyTorch to achieve better results with the Compile API.

> **Note**
>
> Compute capability is a score assigned by NVIDIA to its GPUs. The higher the compute capability, the higher the computing power that GPU provides. PyTorch's official documentation states that the Compile API yields better results on GPUs with compute capability equal to or higher than 8.0.

The following piece of code shows how to load and enable the DenseNet121 model for training:

```
from torchvision import models
device = "cuda"
weights = models.DenseNet121_Weights.DEFAULT
net = models.densenet121(weights=weights)
net.to(device)
net.train()
```

The usage of the Compile API, in this case, is almost the same as before, except for one slight change in the compile line:

```
model = torch.compile(net, mode="reduce-overhead")
```

As you can see, we are invoking a different compiling mode than the one used in the previous example. We did not use the "mode" parameter on the **CNNxFashion-MNIST** case, so the compile function employed the default compiling mode. The compiling mode changes the entire workflow's behavior, allowing us to adjust the generated code so that it fits a particular scenario or requirement.

Figure 3.7 shows the three possible compiling modes:

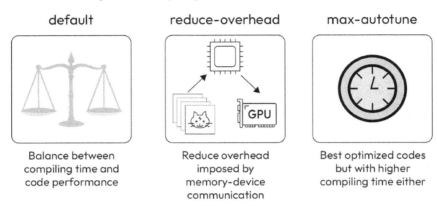

Figure 3.7 – Compiling modes

Here's a breakdown:

- `default`: A balance between compiling time and model performance. As its name suggests, this is the default compiling mode of the function. This option is likely to provide good results in many cases.

- `reduce-overhead`: This is suitable for small batches – which is our present case. This mode reduces the overhead of loading the batch sample to memory and further executes the forward and backward phases on the computing device.

- `max-autotune`: The most optimized code possible. The compiler takes all the time it needs to yield the best-optimized code to run on the target machine or device. Consequently, the time required to compile the model is longer than other modes, which can make this option unfeasible in many practical cases. Even so, this mode is still interesting for experimental purposes since we can evaluate and understand which characteristics make this model better than others generated with default and reduce-overhead modes.

After running the training phase of the eager and compiled models, we got the results listed in *Table 3.1*:

	Eager Model	**Compiled Model**
Overall training time (s)	2,264	1,443
First epoch execution time (s)	47	146

	Eager Model	Compiled Model
Median epoch execution time (s)	45	26
Accuracy (%)	74.26	74.38

Table 3.1 – Results of training the eager and compiled models

The results show that we trained the compiled model 57% faster than its eager version. As expected, the first epoch took much more time to execute on the compiled version than eager mode because the compiling process is performed at that instant. On the other hand, the execution time of the remaining epochs falls from 45 to 26 in the median, representing an execution around 1.73 times faster. Notice that we got this performance improvement without sacrificing the model's quality since both models achieved the same accuracy.

The Compile API accelerated the training phase of the DenseNet121xCIFAR-10 case by almost 60%. But why couldn't this capability do the same for the CNNxFashion-MNIST example? Essentially, the answer lies in two issues: computational burden and computing resources. Let's go over each:

- **Computational burden**: The DenseNet121 model has 7,978,856 parameters. Compared to the 1,630,090 weights of our CNN model, the former is nearly four times greater than the latter. In addition, the dimension of one resized sample of the CIFAR-10 dataset is equal to 244x244x3, which is significantly higher than the dimension of one Fashion-MNIST sample. As discussed in *Chapter 1*, Deconstructing the Training Process model complexity talks directly with the computational burden of the training phase. With a high computational burden, we have a more robust opportunity to accelerate the training phase. Otherwise, it is like removing a speck of dust from a shining surface; there is nothing to do.

- **Computing resources**: We executed our previous experiment in a CPU environment. However, as stated in PyTorch's official documentation, the Compile API tends to provide better results when executed in GPU devices endowed with compute capability higher than 8.0. This is precisely what we did in the DenseNet121xCIFAR-10 case – that is, the training process was executed in a GPU Nvidia A100.

In short, the DenseNet121xCIFAR-10 case, when trained with A100, is a perfect match between computational burden and computing resources. Such a good fit is the key to improving performance through the Compile API.

Now that you're convinced it is a good idea to incorporate this resource in your performance acceleration toolkit, let's understand how the Compile API works behind the scenes.

How does the Compile API work under the hood?

The Compile API is exactly what its name suggests: it is an entry point to access a set of functionalities PyTorch provides to move from eager to graph execution mode. Besides intermediary components and processes, we also have the compiler, which is an entity that's responsible for getting the final work done. There are half a dozen compilers available, each one specialized in generating optimized code for a given architecture or device.

The following sections describe the steps that are involved in the compiling process and the components that make all this possible.

Compiling workflow and components

At this point, we can imagine that the compiling process is much more complex than calling a single line in our code. To transform an eager model into a compiled model, the Compile API executes three steps, namely graph acquisition, graph lowering, and graph compilation, as depicted in *Figure 3.8*:

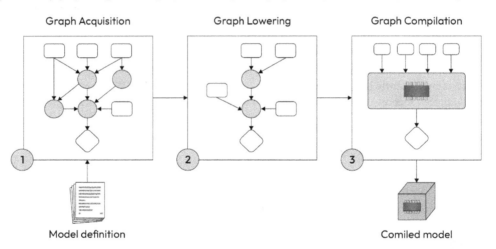

Figure 3.8 – Compiling workflow

Let's discuss each step:

1. **Graph acquisition**: The first step of the compiling workflow, graph acquisition is responsible for capturing model definition and transforming it into a representative graph of primitive operations that are executed on forward and backward phases.

2. **Graph lowering**: With the graph representation at hand, it is time to simplify and optimize the process by fusing, combining, and reducing operations. The simpler the graph is, the lower the time to execute it.

3. **Graph compilation**: The last step consists of generating code for a given target device, such as the CPUs and GPUs of different vendors and architectures, or even for another kind of device, such as a **tensor processing unit** (**TPU**).

PyTorch relies on two main components to execute these steps. **TorchDynamo** executes graph acquisition, whereas the **backend compiler** does graph lowering and compilation, as shown in *Figure 3.9*:

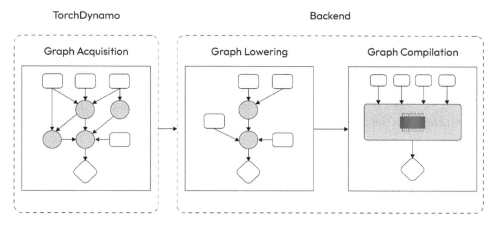

Figure 3.9 – Components of a compiling workflow

TorchDynamo performs graph acquisition by using a new functionality implemented in CPython. This capability is called the Frame Evaluation API and is defined on the **PEP 523**. In short, TorchDynamo captures Python bytecode right before its execution to create a graph representation of operations that are executed by that function or model.

> **Note**
>
> **PEP** stands for **Python Enhancement Proposal**. This document informs the Python community about new features, relevant changes, and general guidance for writing Python code.

After that, TorchDynamo calls the compiler backend, which is the component that's responsible for effectively transforming the graph into a piece of code that can run on the hardware platform. The compiler backend executes the graph lowering and graph compilation steps of the compiling workflow. We'll cover this component in more detail in the next subsection.

Backends

There is a half-dozen backend compilers available to use with the Compile API. The default backend compiler of PyTorch is **TorchInductor**, which generates optimized code for the CPU and GPU through the OpenMP framework and Triton compiler, respectively.

> **Note**
>
> The complete code shown in this section is available at `https://github.com/PacktPublishing/Accelerate-Model-Training-with-PyTorch-2.X/blob/main/code/chapter03/backends.ipynb`

To specify the compiler backend, we must set the parameter backend in the `torch.compile` function. If that parameter is suppressed, the Compile API will use TorchInductor. The following line selects `cudagraphs` as the compiler backend:

```
model = torch.compile(net, backend="cudagraphs")
```

We can easily discover the supported backends of a given environment by running the following command:

```
torch._dynamo.list_backends()

# available backends
['aot_ts_nvfuser',
 'cudagraphs',
 'inductor',
 'ipex',
 'nvprims_nvfuser',
 'onnxrt',
 'tvm']
```

This list shows seven available backends in the environment that's used in our experiments. Remark that backends returned by `list_backends()`, though supported by the current PyTorch installation, are not necessarily ready to be used. This happens because some backends could require additional modules, packages, and libraries to execute.

Of the seven backends available in our environment, only three of them were promptly able to run. *Table 3.2* shows the results that were achieved when we tested the DenseNet121xCIFAR-10 case and compiled it with `aot_ts_nvfuser`, `cudagraphs`, and `inductor`:

	aot_ts_nvfuser	cudagraphs	inductor
Overall training time (s)	2,474	2,290	1,407
First epoch execution time (s)	142	86	140
Median epoch execution time (s)	46	44	25
Accuracy (%)	74.68	77.57	79.90

Table 3.2 – Results of the different backend compilers

The results show that TorchInductor overcame other backends since it executed the training phase 63% faster. Although TorchInductor presented the best result for this case and scenario, it is always interesting to test all the backends that are available in the environment. Furthermore, some backends, such as onnxrt and tvm, specialize in generating models that are suitable for inference.

The next section provides a couple of questions to help you retain what you have learned in this chapter.

Quiz time!

Let's review what we have learned in this chapter by answering a few questions. Initially, try to answer these questions without consulting the material.

> **Note**
>
> The answers to all these questions are available at https://github.com/ PacktPublishing/Accelerate-Model-Training-with-PyTorch-2.X/ blob/main/quiz/chapter03-answers.md.

Before starting this quiz, remember that this is not a test! This section aims to complement your learning process by revising and consolidating the content covered in this chapter.

Choose the correct options for the following questions:

1. Which are the two execution modes of PyTorch?

    A. Horizontal and vertical modes.

    B. Eager and graph modes.

    C. Eager and distributed modes.

    D. Eager and auto modes.

2. In which execution mode does PyTorch execute operations as soon as they appear in the code?

    A. Graph mode.

    B. Eager mode.

    C. Distributed mode.

    D. Auto mode.

3. In which execution mode does PyTorch evaluate the complete set of operations seeking optimization opportunities?

    A. Graph mode.

    B. Eager mode.

  C. Distributed mode.

  D. Auto mode.

4. Compiling a model with PyTorch means changing from eager to graph mode when executing in which of the following phases of the training process?

  A. Forward and optimization.

  B. Forward and loss calculation.

  C. Forward and backward.

  D. Forward and training.

5. Concerning the time to execute the first training epoch in both eager and graph modes, what can we assert?

  A. The time to execute the first training epoch is always the same in both eager and graph modes.

  B. The time to execute the first training epoch in graph mode is always smaller than executing in the eager mode.

  C. The time to execute the first training epoch in graph mode is likely to be higher than executing in the eager mode.

  D. The time to execute the first training epoch in eager mode is likely to be higher than executing in the eager mode.

6. Which phases comprise the compiling workflow that's executed by the Compile API?

  A. Graph forward, graph backward, and graph compilation.

  B. Graph acquisition, graph backward, and graph compilation.

  C. Graph acquisition, graph lowering, and graph optimization.

  D. Graph acquisition, graph lowering, and graph compilation.

7. TorchDynamo is a component of the Compile API that executes which phase?

  A. Graph backward.

  B. Graph acquisition.

  C. Graph lowering.

  D. Graph optimization.

8. TorchInductor is the default compiler backend of PyTorch's Compile API. Which are the other compiler backends?

    A. OpenMP and NCCL.

    B. OpenMP and Triton.

    C. Cudagraphs and IPEX.

    D. TorchDynamo and Cudagraphs.

Now, let's summarize the key takeaways from this chapter.

Summary

In this chapter, you learned about the Compile API, a novel capability that was launched in PyTorch 2.0 and is useful to compile a model – that is, changing the operating mode from eager to graph mode. Models that execute in graph mode tend to train faster, especially in certain hardware platforms. To use the Compile API, we just need to add a single line to our original code. So, it is a simple and powerful technique to accelerate the training process of our models.

In the following chapter, you will learn how to install and configure specialized libraries such as OpenMP and IPEX to speed up the training process of our models.

4

Using Specialized Libraries

Nobody needs to do all things by themselves. Neither does PyTorch! We already know PyTorch is one of the most powerful frameworks for building deep learning models. However, as many other tasks are involved in the model-building process, PyTorch relies on specialized libraries and tools to get the job done.

In this chapter, we will learn how to install, use, and configure libraries to optimize CPU-based training and multithreading.

More important than learning the technical nuances presented in this chapter is catching the message it brings: we can improve performance by using and configuring third-party libraries specialized in tasks that PyTorch relies on. In this sense, we can search for many other options than the ones described in this book.

Here is what you will learn as part of this chapter:

- Understanding the concept of multithreading with OpenMP
- Learning how to use and configure OpenMP
- Understanding IPEX – an API for optimizing the usage of PyTorch on Intel processors
- Understanding how to install and use IPEX

Technical requirements

You can find the complete code of the examples mentioned in this chapter in the book's GitHub repository at `https://github.com/PacktPublishing/Accelerate-Model-Training-with-PyTorch-2.X/blob/main`.

You can access your favorite environment to execute this notebook, such as Google Colab or Kaggle.

Multithreading with OpenMP

OpenMP is a library used for parallelizing tasks by harnessing all the power of multicore processors by using the multithreading technique. In the context of PyTorch, OpenMP is employed to parallelize operations executed in the training phase and to accelerate preprocessing tasks related to data augmentation, normalization, and so forth.

As multithreading is a key concept here, to see how OpenMP works, follow me to the next section to understand this technique.

What is multithreading?

Multithreading is a technique to **parallelize tasks** in a multicore system, which, in turn, is a computer system endowed with **multicore processors**. Nowadays, any computing system has multicore processors; smartphones, notebooks, and even TVs have CPUs with more than one processing core.

As an example, let's look at the notebook that I'm using right now to write this book. My notebook possesses one Intel i5-8265U processor, which has eight cores, as illustrated in *Figure 4.1*:

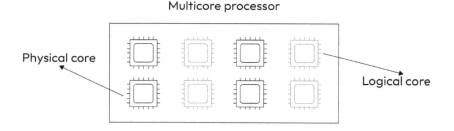

Figure 4.1 – Physical and logical cores

Modern processors have physical and logical cores. A **physical core** is a complete and individual processing unit able to perform any computation. A **logical core** is a processing entity instantiated from the idle resources of physical cores. Therefore, physical cores deliver better performance than logical cores. Thus, we should always prefer to use physical units rather than logical ones.

Nevertheless, from the operating system's point of view, there is no difference between physical and logical cores (i.e., the operating system sees the total number of cores, regardless of whether they are physical or logical).

> **Important note**
> The technology responsible for providing logical cores is called **simultaneous multithreading**. Each vendor has a commercial name for this technology. Intel calls it **hyperthreading**, for example. Details about this topic are out of the scope of this book.

We can inspect details about the processor by using the lscpu command for Linux:

```
[root@laptop] lscpu
Architecture: x86_64
  CPU op-mode(s): 32-bit, 64-bit
  Address sizes: 39 bits physical, 48 bits virtual
  Byte Order: Little Endian
CPU(s): 8
  On-line CPU(s) list: 0-7
Vendor ID: GenuineIntel
  BIOS Vendor ID: Intel(R) Corporation
  Model name: Intel(R) Core(TM) i5-8265U CPU @ 1.60GHz
    BIOS CPU family: 205
    CPU family: 6
    Model: 142
    Thread(s) per core: 2
    Core(s) per socket: 4
    Socket(s): 1
    Stepping: 12
    CPU(s) scaling MHz:   79%
    CPU max MHz: 3900,0000
    CPU min MHz: 400,0000
    BogoMIPS: 3600.00
```

The output shows a bunch of information about the processor, such as the number of cores and sockets, frequency, architecture, vendor name, and so on. Let's examine the most relevant fields to our case:

- **CPU(s)**: The total number of physical and logical cores available on the system. "CPU" is used here as a synonym for "core."

- **On-line CPU(s) list**: The identification of cores available on the system.

- **Socket(s)**: The number of multicore processors installed on the system. In this example, we have just one multicore processor; hence, the number of sockets is 1.

- **Thread(s) per core**: This indicates whether logical cores are enabled in the system. If it has the number 1, there are only physical cores on the system. Otherwise, the system has both physical and logical cores.

- **Core(s) per socket**: The number of physical cores available on each multicore processor.

> **Important note**
> We can use lscpu to get the number of physical and logical cores available on the hardware that we are running on. As you will see in the next sections, this information is essential to optimize the usage of OpenMP.

Modern servers have hundreds of cores. With such computing power on the table, we must find a way to use it properly. Here is where multithreading comes in!

The **multithreading** technique concerns creating and controlling a set of threads to co-operate and accomplish a given task. These threads are spread out on processor cores so that the running program can use different cores to treat distinct pieces of the computing task. As a result, multiple cores work simultaneously on the same task to accelerate its completion.

A **thread** is an operating system entity created by processes. A set of threads created by a given process share the same memory address space. Consequently, threads can communicate with themselves much more easily than processes; they just need to read or write the content of some memory address. On the other hand, processes must resort to more complicated methods such as message exchanging, signals, queues, and so on. This is the reason why we prefer to use threads to parallelize a task instead of processes:

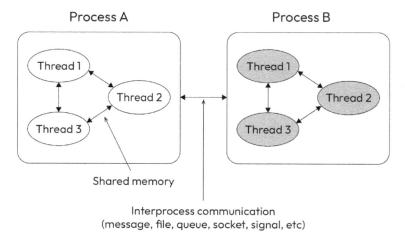

Figure 4.2 – Threads and processes

However, the benefits of using threads come at a price: we must take care of our threads. As threads communicate with themselves through shared memory, they can fall under race conditions when multiple threads intend to write on the same memory region. In addition, the programmer must keep threads synchronized to prevent a thread from waiting indefinitely for some result or action of another thread.

> **Important note**
>
> If the concept of threads and processes is new to you, take a break and watch the following video on YouTube before moving on to the next section: `https://youtu.be/Dhf-DYO1K78`. If you require more profound material, you can read the article written by Roderick Bauer, available at `https://medium.com/@rodbauer/understanding-programs-processes-and-threads-fd9fdede4d88`.

In short, it is hard work to program threads manually (i.e., on our own). However, luckily, OpenMP is here to help. So, let's learn how to use it, along with PyTorch, to accelerate the training phase of our machine learning models.

Using and configuring OpenMP

OpenMP is a framework that is able to encapsulate and abstract many drawbacks related to programming multiple threads. With this framework, we can parallelize our sequential code by employing a set of functions and primitives. When talking about multithreading, OpenMP is the de facto standard. This explains why PyTorch relies on OpenMP as the default backend to parallelize tasks.

Strictly speaking, we do not need to change anything in PyTorch's code to use OpenMP. Nevertheless, there are some configuration tricks that can increase the performance of the training process. Let's see it in practice!

> **Important note**
>
> The complete code shown in this section is available at `https://github.com/PacktPublishing/Accelerate-Model-Training-with-PyTorch-2.X/blob/main/code/chapter04/baseline-cnn_cifar10.ipynb` and `https://github.com/PacktPublishing/Accelerate-Model-Training-with-PyTorch-2.X/blob/main/code/chapter04/gomp-cnn_cifar10.ipynb`.

At first, we will run the same code presented in *Chapter 2, Training Models Faster,* to train the CNN model with the CIFAR-10 dataset. The environment is configured with GNU OpenMP 4.5 and possesses an Intel processor with 32 cores in total, half physical and half logical.

To check the OpenMP version and number of threads used in the current environment, we can execute the `torch.__config__.parallel_info()` function:

```
ATen/Parallel:
    at::get_num_threads() : 16
    at::get_num_interop_threads() : 16
OpenMP 201511 (a.k.a. OpenMP 4.5)
    omp_get_max_threads() : 16
Intel(R) oneAPI Math Kernel Library Version 2022.2
```

```
    mkl_get_max_threads() : 16
Intel(R) MKL-DNN v2.7.3
std::thread::hardware_concurrency() : 32
Environment variables:
    OMP_NUM_THREADS : [not set]
    MKL_NUM_THREADS : [not set]
ATen parallel backend: OpenMP
```

The last line of the output confirms that OpenMP is the parallel backend configured in the environment. We can also see it is OpenMP version 4.5, as well as the number of threads set and values configured for two environment variables. The hardware_concurrency() field shows a value of 32, indicating that the environment is able to run up to 32 threads since the system has 32 cores at maximum.

In addition, the output provides information on the get_num_threads() field, which is the number of threads used by OpenMP. The default behavior of OpenMP is to use a number of threads equivalent to the number of physical cores. So, in this case, the default number of threads is 16.

The training phase took 178 seconds to run 10 epochs. During the training process, we can use the htop command to verify how OpenMP binds threads to cores. In our experiment, PyTorch/OpenMP has made a configuration that is pictorially described in *Figure 4.3*:

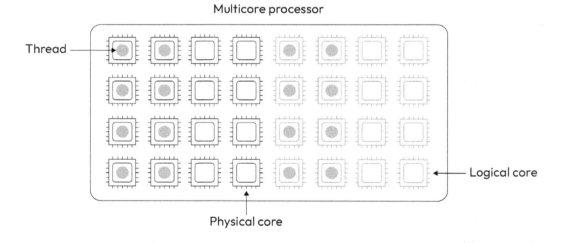

Figure 4.3 – Default OpenMP threads allocation

OpenMP has allocated the set of 16 threads to 8 physical cores and 8 logical cores. As stated in the previous section, physical cores provide better performance than logical ones. Even with physical cores available, OpenMP has used logical cores to execute half of PyTorch's threads.

At first sight, the decision to use logical cores, even when having physical ones available, may sound silly. However, we should remember that processors are used by the entire computing system – that is, they are used for other tasks besides our training process. Therefore, the operating system, together with OpenMP, should try to be fair enough with all demanding tasks – that is, they should also offer the chance to use physical cores.

Despite the default behavior of OpenMP, we can set up a couple of environment variables to change the way OpenMP allocates, controls, and governs threads. The following piece of code, appended to the beginning of our CNN/CIFAR-10 code, modifies OpenMP operations to improve performance:

```
os.environ['OMP_NUM_THREADS'] = "16"
os.environ['OMP_PROC_BIND'] = "TRUE"
os.environ['OMP_SCHEDULE'] = "STATIC"
os.environ['GOMP_CPU_AFFINITY'] = "0-15"
```

These lines set up four environment variables directly from Python code. Before explaining what these variables mean, let's first see the performance improvement they have provided.

The training time of the CNN model using the CIFAR-10 dataset was reduced from 178 seconds to 114 seconds, revealing a performance improvement of 56%! Nothing else was changed in the code! In this execution, OpenMP has created the threads assignment pictorially described in *Figure 4.4*:

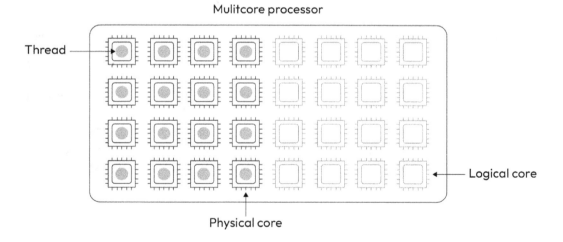

Figure 4.4 – Optimized OpenMP thread allocation

As you can see in *Figure 4.4*, OpenMP has used all 16 physical cores, leaving out the logical cores. We can say that binding threads to physical cores is the primary reason for such an increase in performance.

Let's break down the set of environment variables configured in this experiment to understand how they contribute to improving the performance of our training process:

- `OMP_NUM_THREADS`: This defines the number of threads used by OpenMP. We have set the number of threads to `16`, which is exactly the same value set as the default by OpenMP. Although this configuration did not bring any changes to our scenario, it is essential to know this option to control the number of threads used by OpenMP. This is especially important when running more than one training process simultaneously on the same server.

- `OMP_PROC_BIND`: This determines the thread affinity policy. When set to `TRUE`, this configuration tells OpenMP to keep threads running on the same core during the entire execution. This configuration prevents threads from being moved from cores, thus minimizing performance issues, such as cache missing.

- `OMP_SCHEDULE`: This defines the scheduling policy. As we want to statically bind threads to cores, we should set this variable to a static policy.

- `GOMP_CPU_AFFINITY`: This indicates the cores or processors to be used by OpenMP to execute threads. In order to only use physical cores, we should indicate processor identifications corresponding to the physical cores in the system.

The combination of those variables has greatly accelerated the training process of our CNN model. In short, we forced OpenMP to use only physical cores and keep threads running on the same core they were initially assigned. As a result, we have harnessed all the computing power of the physical cores while minimizing the performance pitfalls from the overhead caused by frequent context switching.

> **Important note**
> Essentially, context switching occurs when the operating system decides to interrupt the execution of a process to give the opportunity of using a CPU to another process.

OpenMP has a couple more variables to control its behavior besides the ones presented in this chapter. To check the current OpenMP configuration, we can set the `OMP_DISPLAY_ENV` environment variable to `TRUE` when running our PyTorch code:

```
OPENMP DISPLAY ENVIRONMENT BEGIN
  _OPENMP = '201511'
  OMP_DYNAMIC = 'FALSE'
  OMP_NESTED = 'FALSE'
  OMP_NUM_THREADS = '16'
  OMP_SCHEDULE = 'STATIC'
  OMP_PROC_BIND = 'TRUE'
  OMP_PLACES = '{0},{1},{2},{3},{4},{5},{6},{7},{8},{9},{10},{11},{12}
               ,{13},{14},{15}'
  OMP_STACKSIZE = '36668818'
```

```
    OMP_WAIT_POLICY = 'PASSIVE'
    OMP_THREAD_LIMIT = '4294967295'
    OMP_MAX_ACTIVE_LEVELS = '2147483647'
    OMP_CANCELLATION = 'FALSE'
    OMP_DEFAULT_DEVICE = '0'
    OMP_MAX_TASK_PRIORITY = '0'
  OPENMP DISPLAY ENVIRONMENT END
```

It is interesting to learn how each of those environment variable changes OpenMP operation; thus, we can fine-tune for particular scenarios. This output is also useful to verify whether changes to the environment variables did indeed take place.

The experiments described in this section used GNU OpenMP since it is the default parallel backend adopted by PyTorch. However, as OpenMP is actually a framework, we have other implementations of OpenMP besides the one provided by GNU. One of those implementations is Intel OpenMP, which is suitable for Intel processor environments.

However, does Intel OpenMP bring relevant improvements? Is it worth using it in place of GNU implementation? See for yourself in the next section!

Using and configuring Intel OpenMP

Intel has its own OpenMP implementation, which promises to deliver better performance in Intel-based environments. As PyTorch comes with GNU implementation by default, we need to take three steps in order to use Intel OpenMP in place of the GNU version:

1. Install Intel OpenMP.
2. Load the Intel OpenMP libraries.
3. Set up specific environment variables for Intel OpenMP.

> **Important note**
> The complete code shown in this section is available at `https://github.com/PacktPublishing/Accelerate-Model-Training-with-PyTorch-2.X/blob/main/code/chapter04/iomp-cnn_cifar10.ipynb`.

The first step is the easiest one. When considering a Python environment based on Anaconda or supporting PIP, we just need to execute one of these commands to install Intel OpenMP:

```
pip install intel-openmp
conda install intel-openmp
```

After installation, we should prioritize loading Intel OpenMP libraries instead of implementing GNU. Otherwise, PyTorch will keep using the libraries of the default OpenMP installation, even with Intel OpenMP installed on the system.

> **Important note**
> If we do not use a PIP or Anaconda-based environment, we can install it on our own. This process requires compiling Intel OpenMP to further install it in the environment.

We enact this configuration by setting the LD_PRELOAD environment variable before running our code:

```
export LD_PRELOAD=/opt/conda/lib/libiomp5.so:$LD_PRELOAD
```

In the environment used for these experiments, the Intel OpenMP library is located at /opt/conda/lib/libiomp5.so. The LD_PRELOAD environment variable allows for forcing the operating system to load libraries before loading the ones configured by default.

At last, we need to set up some environment variables related to Intel OpenMP:

```
import os
os.environ['OMP_NUM_THREADS'] = "16"
os.environ['KMP_AFFINITY'] = "granularity=fine,compact,1,0"
os.environ['KMP_BLOCKTIME'] = "0"
```

OMP_NUM_THREADS has the same meaning as the GNU version, whereas KMP_AFFINITY and KMP_BLOCKTIME are exclusive to Intel OpenMP:

- KMP_AFFINITY: This defines the threads allocation policy. When set to granularity=fine,compact,1,0, Intel OpenMP binds threads to physical cores, besides trying to keep it that way for the entire execution. Thus, in the case of Intel OpenMP, we do not need to pass a list of physical cores to force the usage of physical processors, as we do in GNU implementation.

- KMP_BLOCKTIME: This determines the time that a thread should wait to sleep after completing a task. When set to zero, threads go to sleep immediately after doing their job, thus minimizing the wastage of processor cycles just to wait for another task.

Similar to the GNU version, Intel OpenMP also outputs the current configuration when the OMP_DISPLAY_ENV variable is set to TRUE (shortened output example):

```
OPENMP DISPLAY ENVIRONMENT BEGIN
    _OPENMP='201611'
  [host] OMP_AFFINITY_FORMAT='OMP: pid %P tid %i thread %n bound to OS
                              proc set {%A}'
  [host] OMP_ALLOCATOR='omp_default_mem_alloc'
  [host] OMP_CANCELLATION='FALSE'
```

```
   [host]  OMP_DEBUG='disabled'
   [host]  OMP_DEFAULT_DEVICE='0'
   [host]  OMP_DISPLAY_AFFINITY='FALSE'
   [host]  OMP_DISPLAY_ENV='TRUE'
   [host]  OMP_DYNAMIC='FALSE'
   [host]  OMP_MAX_ACTIVE_LEVELS='1'
   [host]  OMP_MAX_TASK_PRIORITY='0'
   [host]  OMP_NESTED: deprecated; max-active-levels-var=1
   [host]  OMP_NUM_TEAMS='0'
   [host]  OMP_NUM_THREADS='16'
 OPENMP DISPLAY ENVIRONMENT END
```

To compare the performance brought by Intel OpenMP, we take the result provided by the GNU implementation as a baseline. The training time of the CNN model using the CIFAR-10 dataset was reduced from 114 seconds to 102 seconds, resulting in a performance improvement of around 11%. Even though this is not as impressive as the first experiment, the performance gain is still interesting. In addition, note that we can get better results by using other models, datasets, and computing environments.

To summarize, we executed the training process almost 1.7 times faster with the configurations shown in this section. No code modification was necessary to achieve such improvement; only direct configurations were applied at the environment level.

In the next section, we will learn how to install and use an API provided by Intel to accelerate PyTorch's execution on its processors.

Optimizing Intel CPU with IPEX

IPEX stands for **Intel extension for PyTorch** and is a set of libraries and tools provided by Intel to accelerate the training and inference of machine learning models.

IPEX is a clear sign by Intel of highlighting the relevance of PyTorch among machine learning frameworks. After all, Intel has invested a lot of energy and resources in designing and maintaining an API specially created for PyTorch.

It is interesting to say that IPEX strongly relies on libraries provided by the Intel oneAPI toolset. oneAPI contains libraries and tools specific for machine learning applications, such as oneDNN, and other ones to accelerate applications, such as oneTBB, in general.

> **Important note**
> The complete code shown in this section is available at `https://github.com/PacktPublishing/Accelerate-Model-Training-with-PyTorch-2.X/blob/main/code/chapter04/baseline-densenet121_cifar10.ipynb` and `https://github.com/PacktPublishing/Accelerate-Model-Training-with-PyTorch-2.X/blob/main/code/chapter04/ipex-densenet121_cifar10.ipynb`.

Let's learn how to install and use IPEX on our PyTorch code.

Using IPEX

IPEX does not come with PyTorch by default; we need to install it. The easiest way to install IPEX is by using PIP along the same lines we followed for OpenMP in the last section. So, to install IPEX on a PIP environment, we just need to execute the following command:

```
pip install intel_extension_for_pytorch
```

After installing IPEX, we can proceed to the default installation of PyTorch. Once IPEX is available, we are ready to incorporate it into our PyTorch code. The first step concerns importing the IPEX module:

```
import intel_extension_for_pytorch as ipex
```

Using IPEX is very simple. We just need to wrap our model and optimizer with the `ipex.optimize` function and let IPEX do the rest. The `ipex.optimize` function returns an optimized version of the model and optimizer (SGD, Adam, and so on) used to train the model.

To see the performance improvement provided by IPEX, let's test it with the DenseNet121 model and the CIFAR-10 dataset (we introduced both of these in previous chapters).

Our baseline execution concerns training DenseNet121 with the CIFAR-10 dataset over 10 epochs. For the sake of fairness, we have used Intel OpenMP since we are using an Intel-based environment. However, in this case, we do not change the `KMP_BLOCKTIME` parameter:

```
import os
os.environ['OMP_NUM_THREADS'] = "16"
os.environ['KMP_AFFINITY'] = "granularity=fine,compact,1,0"
```

The baseline execution took 1,318 seconds to complete 10 epochs, and the resultant model obtained an accuracy of approximately 70%.

As stated before, using IPEX is very simple; we just need to add one single line to the baseline code:

```
model, optimizer = ipex.optimize(model, optimizer=optimizer)
```

Although `ipex.optimize` accepts other parameters, calling it in this way is usually enough to get what we need.

Our IPEX code took 946 seconds to execute the training process of the DenseNet121 model, representing a performance improvement of nearly 40%. Except for the environment variables configured at the beginning of the code and the usage of that single line, nothing else was changed in the original code. Thus, IPEX accelerated the training process with just one simple modification.

At first sight, IPEX seems similar to the Compile API that we learned about in *Chapter 3, Compiling the Model*. Both of them require adding a single line of code and using the concept of code wrapping. However, the similarities stop there! Unlike the Compile API, IPEX does not compile the model; it replaces some default PyTorch operations with its own implementation.

Follow me to the next section to understand how IPEX works under the hood.

How does IPEX work under the hood?

To understand how IPEX works under the hood, let's profile the baseline code to check which operations the training process has used. The following output shows the 10 most consuming operations executed by the training process:

```
aten::convolution_backward: 27.01%
aten::mkldnn_convolution: 12.44%
aten::native_batch_norm_backward: 8.85%
aten::native_batch_norm: 7.06%
Optimizer.step#Adam.step: 6.80%
aten::add_: 3.75%
aten::threshold_backward: 3.21%
aten::add: 2.75%
aten::mul_: 2.45%
aten::div: 2.19%
```

Our baseline code for DenseNet121 and CIFAR-10 executed those operations commonly employed on convolutional neural networks, such as convolution_backward. No surprise here.

Let's see the profiling output of the IPEX code to verify what changes IPEX has made to our baseline code:

```
torch_ipex::convolution_backward: 32.76%
torch_ipex::convolution_forward_impl: 13.22%
aten::native_batch_norm: 7.98%
aten::native_batch_norm_backward: 7.08%
aten::threshold_backward: 4.40%
aten::add_: 3.92%
torch_ipex::adam_fused_step: 3.14%
torch_ipex::cat_out_cpu: 2.39%
aten::empty: 2.18%
aten::clamp_min_: 1.48%
```

The first thing to notice is the new prefix used on some operations. Besides aten, which denotes the default PyTorch operations library, we also have the torch_ipex prefix. The torch_ipex prefix indicates the operations provided by IPEX. For example, the baseline code used the convolution_backward operation provided by aten, whereas the optimized code used the operation provided by IPEX.

As you can observe, IPEX did not replace every single operation since it does not have an optimized version of all aten operations. This behavior is expected because some operations are already in their most optimized form. In this case, it does not make any sense to try to optimize what is already optimized.

Figure 4.5 summarizes the difference between the default PyTorch code and the optimized versions by IPEX and the Compile API:

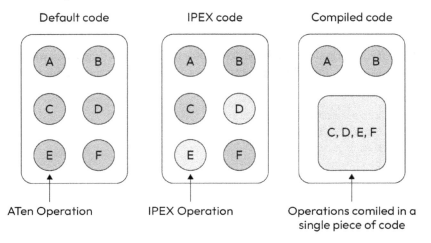

Figure 4.5 – Differences between the default and optimized code generated by IPEX and the Compile API

Unlike the Compile API, IPEX does not create a monolithic piece of compiled code. As a result, the optimization process via ipex.optimize execution is much faster. On the other hand, the compiled code tends to deliver better performance, as we discussed in detail in *Chapter 3, Compiling the Model*.

> **Important note**
> It is interesting to say that we can use IPEX as a compiler backend for the Compile API. In doing this, the torch.compile function will rely on IPEX to compile the model.

As IPEX shows the great gamble made by Intel on PyTorch, it is constantly evolving and receiving frequent updates. Therefore, it is important to use the latest version of this tool to get the newest improvements.

The next section provides some questions to help you retain what you have learned in this chapter.

Quiz time!

Let's review what we have learned in this chapter by answering a few questions. At first, try to answer these questions without consulting the material.

> **Important note**
>
> The answers to all these questions are available at `https://github.com/PacktPublishing/Accelerate-Model-Training-with-PyTorch-2.X/blob/main/quiz/chapter04-answers.md`.

Before starting the quiz, remember that it is not a test at all! This section aims to complement your learning process by revising and consolidating the content covered in this chapter.

Choose the correct option for the following questions.

1. A multicore system can have the following two types of computing cores:

    A. Physical and active.

    B. Physical and digital.

    C. Physical and logical.

    D. Physical and vectorial.

2. A set of threads created by the same process...

    A. May share the same memory address space.

    B. Do not share the same memory address space.

    C. Is impossible in modern systems.

    D. Do share the same memory address space.

3. Which of the following environment variables can be used to set the number of threads used by OpenMP?

    A. OMP_NUM_PROCS.

    B. OMP_NUM_THREADS.

    C. OMP_NUMBER_OF_THREADS.

    D. OMP_N_THREADS.

4. In a multicore system, the usage of OpenMP is able to improve the performance of the training process because it can...

    A. Allocate the process to the main memory.

    B. Bind threads to logical cores.

    C. Bind threads to physical cores.

    D. Avoid the usage of cache memory.

5. Concerning the implementation of OpenMP through Intel and GNU, we can assert that…

    A. There is no difference between the performance obtained by both versions.

    B. The Intel version can outperform GNU's implementation when running on Intel platforms.

    C. The Intel version never outperforms GNU's implementation when running on Intel platforms.

    D. The GNU version is always faster than Intel OpenMP, regardless of the hardware platform.

6. IPEX stands for Intel extension for PyTorch and is defined as…

    A. A set of low-level hardware instructions.

    B. A set of code examples.

    C. A set of libraries and tools.

    D. A set of documents.

7. What is the strategy adopted by IPEX to accelerate the training process?

    A. IPEX enables the usage of special hardware instructions.

    B. IPEX replaces all the training process operations with an optimized version.

    C. IPEX fuses all operations of the training process into a monolithic piece of code.

    D. IPEX replaces some of the default PyTorch operations of the training process with its own optimized implementations.

8. What is necessary to change in our original PyTorch code to use IPEX?

    A. Nothing at all.

    B. We just need to import the IPEX module.

    C. We need to import the IPEX module and wrap the model with the `ipex.optimize()` method.

    D. We just need to use the newest PyTorch version.

Let's summarize what we've covered so far.

Summary

You learned that PyTorch relies on third-party libraries to accelerate the training process. Besides understanding the concept of multithreading, you have learned how to install, configure, and use OpenMP. In addition, you have learned how to install and use IPEX, which is a set of libraries developed by Intel to optimize the training process of PyTorch code executed on Intel-based platforms.

OpenMP can accelerate the training process by employing multiple threads to parallelize the execution of PyTorch code, whereas IPEX is useful for replacing the operations provided by the default PyTorch library by optimizing the operations written specifically for Intel hardware.

In the next chapter, you will learn how to create an efficient data pipeline to keep the GPU working at peak performance during the entire training process.

5

Building an Efficient Data Pipeline

Machine learning is grounded on data. Simply put, the training process feeds the neural network with a bunch of data, such as images, videos, sound, and text. Thus, apart from the training algorithm itself, data loading is an essential part of the entire model-building process.

It turns out that deep learning models deal with huge amounts of data, such as thousands of images and terabytes of text sequences. As a consequence, tasks related to data loading, preparation, and augmentation can severely delay the training process as a whole. So, to overcome a potential bottleneck in the model-building process, we must guarantee an uninterrupted flow of dataset samples to the training process.

In this chapter, we'll explain how to build an efficient data pipeline to keep the training process running smoothly. The main idea is to prevent the training process from being stalled by data-related tasks.

Here is what you will learn as part of this chapter:

- Understanding why it is mandatory to have an efficient data pipeline
- Learning how to increase the number of workers in the data pipeline
- Understanding how to accelerate data transfer through memory pining

Technical requirements

You can find the complete code examples mentioned in this chapter in this book's GitHub repository at https://github.com/PacktPublishing/Accelerate-Model-Training-with-PyTorch-2.X/blob/main.

You can access your favorite environment to execute this notebook, such as Google Colab or Kaggle.

Why do we need an efficient data pipeline?

We'll start this chapter by making you aware of the relevance of having an efficient data pipeline. In the next few subsections, you will understand what a data pipeline is and how it can impact the performance of the training process.

What is a data pipeline?

As you learned in *Chapter 1*, *Deconstructing the Training Process*, the training process is composed of four phases: forward, loss calculation, optimization, and backward. The training algorithm iterates on dataset samples until there's a complete epoch. Nevertheless, there is an additional phase we excluded from that explanation: **data loading**.

The forward phase invokes data loading to get dataset samples to execute the training process. More specifically, the forward phase calls the data loading process on each iteration to get the data required to execute the current training step, as shown in *Figure 5.1*:

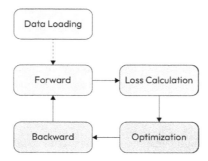

Figure 5.1 – Data loading process

In short, the data loading executes three main tasks:

1. **Loading**: This step involves reading data from a disk and loading it in memory. We can load data into main memory (DRAM) or directly into GPU memory (GRAM).

2. **Preparation**: Usually, we need to prepare data before using it in the training process, such as by performing normalization and resizing.

3. **Augmentation**: When the dataset is small, we must augment it by creating new samples derived from the original ones. Otherwise, the neural network won't be able to catch the intrinsic knowledge presented in the data. Augmentation tasks include rotation, mirroring, and flipping images.

In general, data loading executes those tasks *on demand*. So, when invoked by the forward phase, it starts to execute all tasks to deliver a dataset sample to the training process. Then, we can see this whole process as a **data pipeline**, in which the data is processed before being used to train the neural network.

A data pipeline (pictorially described in *Figure 5.2*) is similar to an industrial production line. The original dataset sample is processed sequentially and transformed until it is ready to feed the training process:

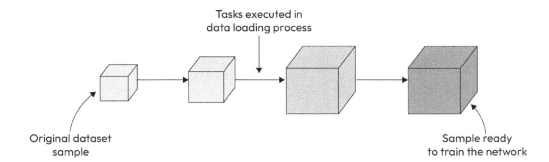

Figure 5.2 – Data pipeline

In many cases, model quality is dependent on transformations that are made to the dataset. This is particularly true for small datasets – for which augmentation is almost mandatory – and datasets comprised of poor-quality images.

In other situations, we do not need to make any modifications to the sample to reach a highly accurate model, perhaps only changing the data format or something like that. In such cases, the data pipeline is limited to loading dataset samples from memory or disk and delivering them to the forward phase.

Regardless of tasks related to transforming, preparing, and converting data, we need to build a data pipeline to feed the forward phase. In PyTorch, we can use components provided by the `torch.utils.data` API to create a data pipeline, as we will see in the next section.

How to build a data pipeline

The `torch.utils.data` API provides two components to build a data pipeline: `Dataset` and `DataLoader` (as shown in *Figure 5.3*). The former is used to indicate the source of the dataset (local files, downloads from the internet, and so on) and to define the set of transformations to be applied to the dataset, whereas the latter is used as an interface to obtain samples from the dataset:

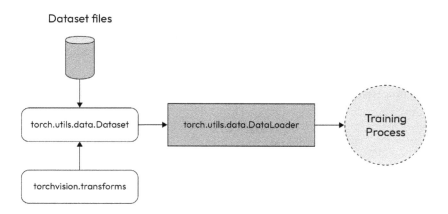

Figure 5.3 – The DataLoader and Dataset components

In practical terms, the training process talks directly to `DataLoader` to consume dataset samples. Thus, the forward phase asks `DataLoader` for a dataset sample on each training step.

The following piece of code shows an example of the basic usage of `DataLoader`:

```
transform = transforms.Compose(transforms.Resize(255))

dataset = datasets.CIFAR10(root=data_dir,
                           train=True,
                           download=True,
                           transform=transform)

dataloader = torch.utils.data.DataLoader(dataset, batch_size=128)
```

This piece of code creates a `DataLoader` instance, namely `dataloader`, to provide samples with batch sizes equal to 128.

Note

Note that `Dataset` was not used directly in this case since CIFAR-10 encapsulates dataset creation.

There are other strategies to build a data pipeline in PyTorch, but `Dataset` and `DataLoader` commonly attend to most cases.

Next, we'll learn how an inefficient data pipeline can slow down the entire training process.

Data pipeline bottleneck

Depending on the complexity of tasks incorporated into the data pipeline, as well as the size of the dataset sample, data loading can take a reasonable time to finish. As a consequence, we can throttle the entire building process.

In general, data loading is executed on the CPU, whereas training takes place on the GPU. As the CPU is much slower than the GPU, the GPU can stay idle, waiting for the next sample to proceed with the training process. The higher the complexity of tasks executed on data feeding, the worse the impact on the training phase.

As shown in *Figure 5.4*, data loading uses the CPU to process dataset samples. When samples become ready, the training phase uses them to train the network. This procedure is continuously executed until all the training steps are completed:

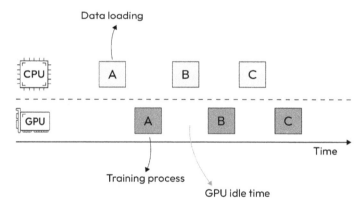

Figure 5.4 – Bottleneck caused by the inefficient data pipeline

Although this procedure seems fine at first sight, we are wasting GPU computing power because it stays idle between training steps. The desired behavior is more like what's shown in *Figure 5.5*:

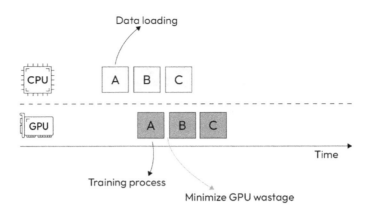

Figure 5.5 – Efficient data pipeline

Unlike the previous scenario, the interleaving time between training steps is hardly reduced since samples are loaded earlier, ready to feed the training process that's executed on the GPU. As a consequence, we experience an overall speedup in the model-building process.

In the next section, we'll learn how to accelerate the data-loading process by making a couple of simple changes to the code.

Accelerating data loading

Accelerating data loading is crucial to get an efficient data pipeline. In general, the following two changes are enough to get the work done:

- Optimizing a data transfer between the CPU and GPU
- Increasing the number of workers in the data pipeline

Putting it that way, these changes may sound tougher to implement than they are. Making these changes is quite simple – we just need to add a couple of parameters when creating the DataLoader instance for the data pipeline. We will cover this in the following subsections.

Optimizing a data transfer to the GPU

To transfer data from main memory to the GPU, and vice versa, the device driver must ask the operating system to pin or lock a portion of memory. After receiving access to that pinned memory, the device driver starts to copy data from the original memory location to the GPU, but using the pinned memory as a **staging area**:

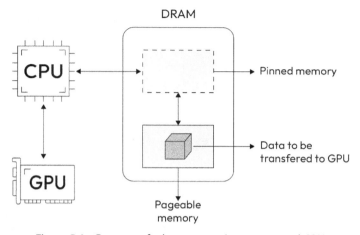

Figure 5.6 – Data transfer between main memory and GPU

The usage of pinned memory in the middle of this process is obligatory because the device driver cannot copy data directly from pageable memory to the GPU. There are architectural issues involved in that procedure, which explains this behavior. Anyway, we can assert that this **double-copy procedure** can negatively affect the performance of the data pipeline.

> **Note**
>
> You can find more information about pinned memory transfer here: `https://developer.nvidia.com/blog/how-optimize-data-transfers-cuda-cc/`.

To overcome this problem, we can tell the device driver to allocate a portion of pinned memory right away instead of requesting a pageable memory area, as usual. By doing so, we can eliminate the unnecessary copy between pageable and pinned memory, thus greatly reducing the overhead involved in GPU data transfer, as shown in *Figure 5.7*:

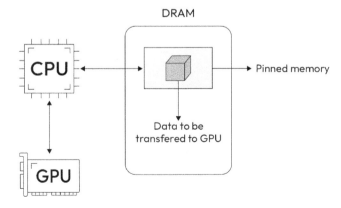

Figure 5.7 – Data transfer using pinned memory

To enable this option on the data pipeline, we need to turn on the `pin_memory` flag while creating `DataLoader`:

```
dataloader = torch.utils.data.DataLoader(dataset,
                                         batch_size=128,
                                         pin_memory=True)
```

Nothing else is necessary. But if it is so simple to implement and highly beneficial, why does PyTorch not enable this feature by default? There are two reasons for this:

- *Request for pinned memory can fail*: As stated on the Nvidia developer blog, "*It is possible for pinned memory allocation to fail, so you should always check for errors.*" Thus, there is no guarantee of success in allocating pinned memory.

- *Increase in memory usage*: Modern operating systems commonly adopt a paging mechanism to manage memory resources. By using this strategy, the operating system can move unused memory pages to disk to free space on main memory. However, pinned memory allocation makes the operating system unable to move pages of that area, disrupting the memory management process and increasing the effective amount of memory usage.

Besides optimizing GPU data transfer, we can configure workers to accelerate data pipeline tasks, as discussed in the next section.

Configuring data pipeline workers

The default operation mode of `DataLoader` uses a **single process** to execute the data pipeline. In other words, a data pipeline has only one process working on loading, preparation, and data augmentation, as shown in *Figure 5.8*. Consequently, `DataLoader` stays idle waiting for samples, wasting valuable computing resources. Such harmful behavior becomes worse in a heavy data pipeline:

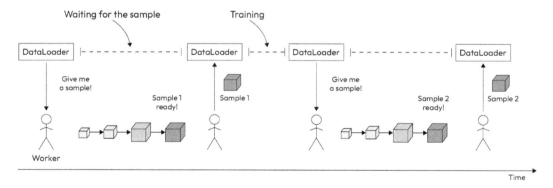

Figure 5.8 – Single worker data pipeline

Fortunately, we can increase the number of processes operating on the data pipeline – that is, we can increase the number of data pipeline *workers*. When set to more than one worker, PyTorch will create additional processes to work simultaneously in more than one dataset sample:

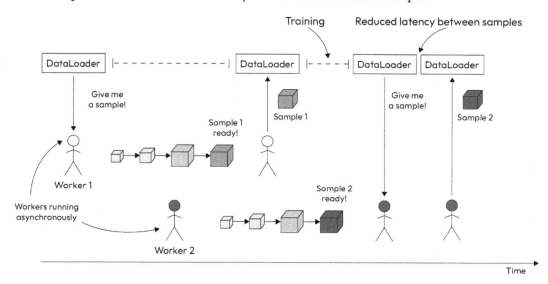

Figure 5.9 – Multi-worker data pipeline

As illustrated in *Figure 5.9*, DataLoader receives **Sample 2** as soon as it asks for a new sample. This happens because **Worker 2** has started to work asynchronously and simultaneously on that sample, even without receiving a request to do it.

To increase the number of workers, we just need to set the num_workers parameter on DataLoader creation:

```
torch.utils.data.DataLoader(train_dataset,
                            batch_size=128,
                            num_workers=2)
```

We'll look at a practical performance improvement case in the next section.

Reaping the rewards

> **Note**
>
> The complete code shown in this section is available at https://github.com/
> PacktPublishing/Accelerate-Model-Training-with-PyTorch-2.X/
> blob/main/code/chapter05/complex_pipeline.ipynb.

To see a relevant performance improvement provided by those changes, we need to apply them to a complex data pipeline – that is, a worthy data pipeline! Otherwise, there is no room for performance gain. Therefore, we will adopt a data pipeline composed of seven tasks as our baseline, as shown here:

```
transform = transforms.Compose(
            [transforms.Resize(255),
             transforms.CenterCrop(size=224),
             transforms.RandomHorizontalFlip(p=0.5),
             transforms.RandomRotation(20),
             transforms.GaussianBlur(kernel_size=3),
             transforms.ToTensor(),
             transforms.Normalize([0.485, 0.456, 0.406],
                                  [0.229, 0.224, 0.225])
            ])
```

For each sample, the data loading process applies five transformations, namely resizing, cropping, flipping, rotation, and Gaussian blur. After applying these transformations, data loading converts the resultant image into a tensor data type. Finally, the data is normalized according to a set of parameters.

To assess performance improvement, we used this pipeline to train the **ResNet121** model over the **CIFAR-10** dataset. The training process, which is comprised of 10 epochs, took 1,892 seconds to complete, even running on an environment endowed with an NVIDIA A100 GPU:

```
Epoch [1/10], Loss: 1.1507, time: 187 seconds
Epoch [2/10], Loss: 0.7243, time: 199 seconds
Epoch [3/10], Loss: 0.4129, time: 186 seconds
Epoch [4/10], Loss: 0.3267, time: 186 seconds
Epoch [5/10], Loss: 0.2949, time: 188 seconds
Epoch [6/10], Loss: 0.1711, time: 186 seconds
Epoch [7/10], Loss: 0.1423, time: 197 seconds
Epoch [8/10], Loss: 0.1835, time: 186 seconds
Epoch [9/10], Loss: 0.1127, time: 186 seconds
Epoch [10/10], Loss: 0.0946, time: 186 seconds
Training time: 1892 seconds
Accuracy of the network on the 10000 test images: 92.5 %
```

Note that this data pipeline is significantly heavier than the ones we've adopted so far in this book, which is exactly what we want!

To use pinned memory and enable multi-worker capability, we must set those two parameters on the original code:

```
torch.utils.data.DataLoader(train_dataset,
                            batch_size=128,
                            pin_memory=True,
                            num_workers=8)
```

After applying these changes to our code, we'll get the following result:

```
Epoch [1/10], Loss: 1.3163, time: 86 seconds
Epoch [2/10], Loss: 0.5258, time: 84 seconds
Epoch [3/10], Loss: 0.3629, time: 84 seconds
Epoch [4/10], Loss: 0.3328, time: 84 seconds
Epoch [5/10], Loss: 0.2507, time: 84 seconds
Epoch [6/10], Loss: 0.2655, time: 84 seconds
Epoch [7/10], Loss: 0.2022, time: 84 seconds
Epoch [8/10], Loss: 0.1434, time: 84 seconds
Epoch [9/10], Loss: 0.1462, time: 84 seconds
Epoch [10/10], Loss: 0.1897, time: 84 seconds
Training time: 846 seconds
Accuracy of the network on the 10000 test images: 92.34 %
```

We have reduced the training time from 1,892 to 846 seconds, representing an impressive performance improvement of 123%!

The next section provides a couple of questions to help you retain what you have learned in this chapter.

Quiz time!

Let's review what we have learned in this chapter by answering a few questions. Initially, try to answer these questions without consulting the material.

> **Note**
>
> The answers to all these questions are available at https://github.com/
> PacktPublishing/Accelerate-Model-Training-with-PyTorch-2.X/
> blob/main/quiz/chapter05-answers.md.

Before starting this quiz, remember that this is not a test! This section aims to complement your learning process by revising and consolidating the content covered in this chapter.

Choose the correct options for the following questions:

1. What three main tasks are executed during the data loading process?

    A. Loading, scaling, and resizing.

    B. Scaling, resizing, and loading.

    C. Resizing, loading, and filtering.

    D. Loading, preparation, and augmentation.

2.   Data loading feeds which phase of the training process?

   A. Forward.

   B. Backward.

   C. Optimization.

   D. Loss calculation.

3.   Which components provided by the `torch.utils.data` API can be used to implement a data pipeline?

   A. `Datapipe` and `DataLoader`.

   B. `Dataset` and `DataLoading`.

   C. `Dataset` and `DataLoader`.

   D. `Datapipe` and `DataLoading`.

4.   Besides increasing the number of workers in the data pipeline, what can we do to improve the performance of the data loading process?

   A. Reduce the size of the dataset.

   B. Do not use a GPU.

   C. Avoid the usage of high-dimensional images.

   D. Optimize data transfer between the CPU and GPU.

5.   How can we accelerate the data transfer between the CPU and GPU?

   A. Use smaller datasets.

   B. Use the fastest GPUs.

   C. Allocate and use pinned memory instead of pageable memory.

   D. Increase the amount of main memory.

6.   What should we do to enable the usage of pinned memory on `DataLoader`?

   A. Nothing. It is already enabled by default.

   B. Set the `pin_memory` parameter to `True`.

   C. Set the `experimental_copy` parameter to `True`.

   D. Update PyTorch to version 2.0.

7. Why can using more than one worker on the pipeline accelerate data loading on PyTorch?

    A. PyTorch reduces the amount of allocated memory.

    B. PyTorch enables the usage of special hardware capabilities.

    C. PyTorch uses the fastest links to communicate with GPUs.

    D. PyTorch processes simultaneously more than one dataset sample.

8. Which of the following is true when making a request to allocate pinned memory?

    A. It is always satisfied.

    B. It can fail.

    C. It always fails.

    D. It cannot be done through PyTorch.

Now, let's summarize what we've covered in this chapter.

Summary

In this chapter, you learned that the data pipeline is an important piece of the model-building process. Thus, an efficient data pipeline is essential to keep the training process running without interruptions. Besides optimizing data transfer to the GPU through memory pining, you have learned how to enable and configure a multi-worker data pipeline.

In the next chapter, you will learn how to reduce model complexity to speed up the training process without penalizing model quality.

6
Simplifying the Model

Have you heard about parsimony? **Parsimony**, in the context of model estimation, concerns keeping a model as simple as possible. Such a principle comes from the assumption that complex models (models with a higher number of parameters) overfit the training data, thus reducing the capacity to generalize and make good predictions.

In addition, simplifying neural networks has two main benefits: reducing the model training time and making the model feasible to run in resource-constrained environments. One of the approaches to simplifying a model relies on reducing the number of parameters of the neural network by employing pruning and compression techniques.

In this chapter, we show how to simplify a model by reducing the number of parameters of the neural network without sacrificing its quality.

Here is what you will learn as part of this chapter:

- The key benefits of simplifying a model
- The concept and techniques of model pruning and compression
- How to use the Microsoft NNI toolkit to simplify a model

Technical requirements

You can find the complete code of examples mentioned in this chapter in the book's GitHub repository at `https://github.com/PacktPublishing/Accelerate-Model-Training-with-PyTorch-2.X/blob/main`.

You can access your favorite environments to execute this notebook, such as Google Colab or Kaggle.

Knowing the model simplifying process

In simpler words, simplifying a model concerns removing connections, neurons, or entire layers of the neural network to get a lighter model, i.e., a model with a reduced number of parameters. Naturally, the efficiency of the simplified version must be very close to the one achieved by the original model. Otherwise, simplifying the model does not make any sense.

To understand this topic, we must answer the following questions:

- Why simplify a model? (reason)
- How do we simplify a model? (process)
- When do we simplify a model? (moment)

We will go through each of these questions in the following sections to get an overall understanding of model simplification.

> **Note**
>
> Before moving on in this chapter, it is essential to say that model simplification is still an open research area. Consequently, some concepts and terms cited in this book may differ a little bit from other materials or how they are employed on frameworks and toolkits.

Why simplify a model? (reason)

To get an insight into the reasons why a model should be simplified let's make use of a simple yet nice analogy.

Consider the hypothetical situation where we must build a bridge to connect two sides of a river. For safety purposes, we decided to put a column in every two meters of the bridge, as shown in *Figure 6.1*:

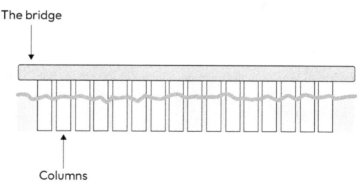

Figure 6.1 – Bridge analogy

The bridge seems pretty safe, being sustained by its 16 columns. However, someone could look at the project and say we do not need all 16 columns to maintain the bridge. As the bridge's designers, we could argue that safety comes first; thus, there is no problem with having additional columns to undoubtedly guarantee the bridge's integrity.

Even so, what if we could spare the columns a little bit without affecting the bridge's structure? To put it differently, perhaps using 16 columns to support the bridge is too much in terms of safety. As we can see in *Figure 6.2*, maybe nine columns could work fine for this scenario:

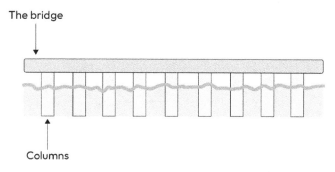

Figure 6.2 – The bridge remains up after removing some columns

If we could put fewer columns in the bridge and keep it as safe as before, we would reduce the budget and building time, as well as simplify the maintenance process in the future. There will not be a reasonable argument to refute this approach.

This naïve analogy is helpful to heat our discussion about the reasons to simplify a model. As for the columns in the bridge, does a neural network model need all of its parameters to achieve good accuracy? The answer is not straightforward and depends on issues such as the problem type and model itself. However, considering that a simplified version of the model performs precisely as the original, why not try the former?

After all, it is undeniable that simplifying a model has clear benefits:

- **Accelerating the training process**: A neural network composed of a lower number of parameters is usually faster to train. As discussed in *Chapter 1, Deconstructing the Training Process,* the number of parameters directly impacts the computational burden of neural networks.

- **Running inference on resource-constrained environments**: Some models are too large to store and so complex to execute that they do not fit in environments comprised of reduced memory and computing capacity. In this case, the only way to run the model in these environments relies on simplifying it as much as possible.

Now that the benefits of simplifying the model are crystal clear let's jump to the next section to learn how to execute this process.

How to simplify a model? (process)

We simplify a model by applying a workflow comprised of two steps: **pruning** and **compression**:

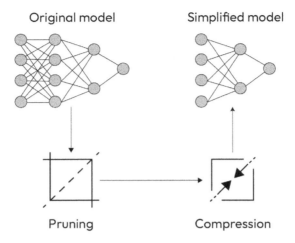

Figure 6.3 – Simplifying the workflow

As pictorially described in *Figure 6.3*, the simplifying workflow receives a dense neural network as input, where all neurons are fully connected to themselves, and this outputs a simplified version of the original model. In other words, the workflow transforms a dense neural network into a sparse neural network.

> **Note**
>
> The terms **dense** and **sparse** come from math and are used to describe matrices. A dense matrix is filled with useful numbers, whereas a sparse matrix possesses a relevant quantity of null values (zeros). As the parameters of neural networks can be expressed in n-dimensional matrices, a non-fully connected neural network is also known as a sparse neural network because of the high number of null connections between neurons.

Let's zoom in on the workflow to understand the role played by each step, starting with the pruning phase.

The pruning phase

The **pruning phase** is responsible for receiving the original model and cutting off the parameters present in the connections (weights), neurons (bias), and filters (kernel values), resulting in a pruned model:

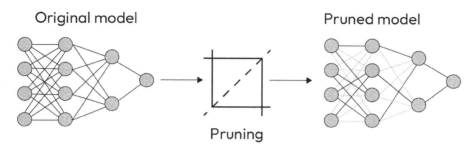

Figure 6.4 – Pruning phase

As shown in *Figure 6.4*, many connections from the original model have been disabled (represented as opaque lines in the pruned model). The pruning phase decides which parameters should be removed depending on the **technique** applied during the process.

A pruning technique has three dimensions:

- **Criterion**: Defines which parameters to cut off

- **Scope**: Determines whether to drop an entire structure (neuron, layer, or filter) or isolated parameters

- **Method**: Defines whether to prune the network at once or iteratively prune the model until it reaches some stop criteria

> **Note**
>
> Model pruning is a brave new world. Therefore, you can easily find many scientific papers proposing new methods and solutions to this area. One exciting paper is entitled *A Survey on Deep Neural Network Pruning: Taxonomy, Comparison, Analysis, and Recommendations*, which summarizes the recent advances in this field and briefly presents other simplifying techniques such as quantization and knowledge distillation.

In practice, the pruned model occupies the same amount of memory and requires the same computational capacity as the original model. This happens because the null parameters, although not having a practical effect on the forward and backward computations and results, were *not definitively removed from the network*.

For example, suppose two fully connected layers comprised of three neurons each. The weights of the connections can be represented as a matrix, as shown in *Figure 6.5*:

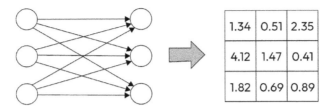

Figure 6.5 – Weights represented as a matrix

After applying the pruning process, the neural network had three connections disabled, as we can see in *Figure 6.6*:

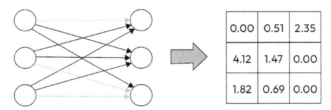

Figure 6.6 – Pruned weights changed to null

Remark that the weights were changed to null (0.00), leaving the connections (represented by those weights) out of the calculations carried out by the network. Therefore, these connections merely do not exist in terms of the meaning of neural network results.

However, the data structure is exactly the same as the original model. We still have nine float numbers, and all of them are still being multiplied (although with no practical effect) by the output of their respective neurons. From the point of view of memory consumption and computational burden, nothing has changed so far.

Well, if the purpose of simplifying a model is to reduce the number of parameters, why do we continue having the same structure as before? Keep calm, and let's execute the second phase of the simplifying workflow: the compression phase.

Compression phase

As illustrated in *Figure 6.7*, the **compression phase** receives a pruned model as input and generates a new brain model comprised only of the non-pruned parameters, i.e., the parameters that the pruning process left untouched:

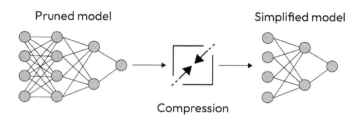

Figure 6.7 – Compression phase

The new network can have a shape utterly different from the original model, with a distinct disposal of neurons and layers. Beyond everything, the compression process is free to generate a new model since it respects the parameters preserved by the pruning step.

Therefore, the compression phase effectively removes the parameters of the pruned model, resulting in a truly simplified model. Let's take the example presented in *Figure 6.8* to understand what happens after model compression:

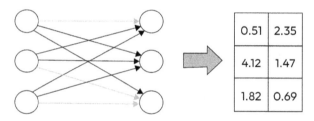

Figure 6.8 – Model compression applied to a pruned network

The set of disabled parameters—connections, in this example—were removed from the model, reducing the weights' matrix by one-third. Consequently, the weights' matrix now occupies less memory and requires fewer operations to complete the forward and backward phases.

> **Note**
>
> We can think about the relationship between pruning and compression phases as the process of deleting a file from a disk. When we ask the operating system to delete a file, it just marks the block as free where the file is allocated. In other words, the file content is still there and will be erased only when overwritten by a new file.

When do we simplify a model? (moment)

We can simplify a model before or after training the neural network. In other words, the model simplification process can be applied to non-trained, pre-trained, or trained models, as explained in the following:

- **Non-trained model**: Our goal here is to speed up the training process. As the model has not been trained yet, the neural network is filled with random parameters, preventing most pruning techniques from doing a useful job. To work around this issue, we usually run a **warmup phase**, which consists of training the network over a single epoch before simplifying the model.

- **Pre-trained model**: We use pre-trained networks, which have a general efficiency on a general domain, to tackle a particular problem of that domain. In this case, we do not need to execute the warmup phase because the model has already been trained.

- **Trained model**: Simplifying a trained model is usually done to deploy the trained network in a resource-constrained environment.

Now that we have the answers to the questions about model simplification, shall we use PyTorch, alongside a toolkit, to simplify a model? Follow me to the next section to learn how to do it!

Using Microsoft NNI to simplify a model

Neural Network Intelligence (**NNI**) is an open-source project created by Microsoft to help deep learning practitioners automate tasks such as hyperparameter automatization and neural architecture searches.

NNI also has a set of tools to deal with model simplification in a simpler and straightforward manner. So, we can easily simplify a model by adding a couple of lines to our original code. NNI supports PyTorch and other well-known deep learning frameworks such as TensorFlow.

> **Note**
>
> PyTorch has its own API to prune models, namely `torch.prune`. Unfortunately, at the time of writing this book, this API does not provide a mechanism to compress a model. Therefore, we have decided to introduce NNI as the solution to accomplish this task. More information about NNI can be found at `https://github.com/microsoft/nni`.

Let's start by getting an overview of NNI in the next section.

Overview of NNI

Because NNI is not a native component of PyTorch, we need to install it via pip by executing the following command:

```
pip install nni
```

NNI has many modules, but for the purpose of model simplification, we are going to use only two of them, namely pruning and speedup:

```
from nni.compression.pytorch import pruning, speedup
```

Pruning module

The pruning module provides a set of pruning techniques, also known as **pruners**. Each pruner applies a specific method to prune the model and requires a particular set of parameters. Among the parameters needed by the pruner, two of them are mandatory: the model and a **configuration list**.

The configuration list is a dictionary-based structure used to control pruner behavior. From the configuration list, we can indicate which structures (layers, operations, or filters) the pruner must work on and which ones it should leave untouched.

For example, the following configuration list tells the pruner to work on all layers implementing the Linear operator (layers created with the class torch.nn.Linear), except the one named layer4. In addition, the pruner will try to nullify 30% of the parameters, as indicated on the sparse_ratio key:

```
config_list = [{
    'op_types': ['Linear'],
    'exclude_op_names': ['layer4'],
    'sparse_ratio': 0.3
}]
```

> **Note**
> You can find the complete list of key-value pairs accepted by the config list at https://nni.readthedocs.io/en/stable/compression/config_list.html.

After setting the configuration list, we have everything to instantiate the pruner, as follows:

```
pruner = pruning.L1NormPruner(model, config_list)
```

The most crucial method provided by the pruner is called compress. Despite what the name suggests, it executes the pruning process by applying the corresponding pruning algorithm.

The compress method returns a data structure called **masks**, which denotes what parameters were dropped by the pruning algorithm. This information is used further to remove the pruned parameters from the network effectively.

> **Note**
>
> As stated before, the simplifying process is still ongoing. Therefore, we can face some tricks, such as the incoherent usage of terms. This is the reason why NNI calls the pruning phase `compress`, though the compressing step is accomplished by another method called `speedup`.

Note that, until this point, nothing has indeed changed to the original model; not yet. To remove the pruned parameters effectively, we must rely on the `speedup` module.

The speedup module

The `speedup` module provides a class named `ModelSpeedup`, which is used to create a **speeder**. A speeder executes the compression phase on the pruned model, i.e., it effectively removes the parameters dropped by the pruner.

Along the lines of pruners, we also must instantiate an object from the `ModelSpeedup` class. This class requires three obligatory parameters: the pruned model, an input sample, and the masks generated by the pruner:

```
speeder = speedup.ModelSpeedup(model, input_sample, masks)
```

After that, we just need to call the `speedup_model` method so the speeder can compress the model and return a simplified version of the original model:

```
speeder.speedup_model()
```

Now that you have an overview of the fundamental steps to simplify a model through NNI let's jump to the next section to learn how to use this toolkit in a practical example.

NNI in action!

To see NNI working in practice let's simplify our well-known CNN model. In this example, we are going to simplify this model by training it over the CIFAR-10 dataset.

> **Note**
>
> The complete code shown in this section is available at `https://github.com/PacktPublishing/Accelerate-Model-Training-with-PyTorch-2.X/blob/main/code/chapter06/nni-cnn_cifar10.ipynb`.

Let's start by counting the original number of parameters of the CNN model:

```
model = CNN()
print(count_parameters(model))
2122186
```

The CNN model has 2,122,186 parameters distributed among the biases, weights, and filters of the neural network. We trained this model using the CIFAR-10 dataset only during 10 epochs since we are interested in comparing the training time and corresponding accuracy between distinct pruning configurations. So, the original model took 122 seconds to train in a CPU-based machine, reaching an accuracy of 47.10%.

Okay, so let's remove some pillars of the bridge to see whether it still stands up. We are going to simplify the CNN model by considering the following strategy:

- Operation types: Conv2d

- Sparsity per layer: 0.50

- Pruner algorithm: L1 Norm

This strategy tells the simplification process to look at only the convolutional layers of the neural network, and, for each layer, the pruning algorithm must throw away 50% of the parameters. As we are simplifying a fresh model, we need to execute a warmup phase to populate the network with some valuable parameters.

For this experiment, we have chosen the L1 Norm pruner, which removes parameters according to the magnitude measured by L1 normalization. In simpler words, the pruner will drop off the parameters, which will have a minor influence on neural network results.

> **Note**
>
> You can find more information about the L1 Norm pruner at https://nni.readthedocs.io/en/stable/reference/compression/pruner.html#l1-norm-pruner.

The following excerpt of code shows a couple of lines needed to simplify the CNN model by applying the aforementioned strategy:

```
config_list = [{'op_types': ['Conv2d'],
                'sparsity_per_layer': 0.50}]
pruner = pruning.L1NormPruner(model, config_list)
_, masks = pruner.compress()
pruner._unwrap_model()
input_sample, _ = next(iter(train_loader))
speeder = speedup.ModelSpeedup(model, input_sample, masks)
speeder.speedup_model()
```

During the simplification process, NNI will output a bunch of lines, such as these:

```
[2023-09-23 19:44:30] start to speedup the model
[2023-09-23 19:44:30] infer module masks...
[2023-09-23 19:44:30] Update mask for layer1.0
```

```
[2023-09-23 19:44:30] Update mask for layer1.1
[2023-09-23 19:44:30] Update mask for layer1.2
[2023-09-23 19:44:30] Update mask for layer2.0
[2023-09-23 19:44:30] Update mask for layer2.1
[2023-09-23 19:44:30] Update mask for layer2.2
[2023-09-23 19:44:30] Update mask for .aten::size.8
[2023-09-23 19:44:30] Update mask for .aten::Int.9
[2023-09-23 19:44:30] Update mask for .aten::reshape.10
```

After the process has been carried out, we can verify that the number of parameters of the original neural network has decreased by around 50%, as expected:

```
print(count_parameters(model))
1059306
```

Well, the model is smaller and simpler. But how about the training time and efficiency? Let's find out!

We trained the simplified model against CIFAR-10 by using the same hyperparameters through the same number of epochs. The training process of the simplified model took 89 seconds to complete, representing a performance improvement of 37%! Although the model's efficiency decreased a little bit (from 47.10% to 42.71%), it remains very close to the original version.

It is interesting to note the trade-off between training time, accuracy, and sparsity ratio. As shown in *Table 6.1*, the model's efficiency falls to 38.87% when 80% of parameters are removed from the network. On the other hand, the training process took only 76 seconds to finish, which is 61% faster than training the original network:

Sparsity per layer	Training time	Accuracy
10%	118	47.26%
20%	113	45.84%
30%	107	44.66%
40%	100	45.18%
50%	89	42.71%
60%	84	41.90%
70%	81	40.84%
80%	76	38.87%

Table 6.1 – Relation between model accuracy, training time, and sparsity level

As the saying goes, there is no free lunch So, accuracy is expected to deteriorate slightly when simplifying the model. The goal here is to find a balance between a tolerable decrease in the model's quality in the face of a reasonable performance improvement.

In this section, we have learned how to use NNI to simplify our model. By changing a couple of lines on our original code, we can simplify the model by cutting off a certain number of connections, thus contributing to reducing the training time, yet retaining the model's quality.

The next section brings a couple of questions to help you retain what you have learned in this chapter.

Quiz time!

Let's review what we have learned in this chapter by answering a few questions. At first, try to answer these questions without consulting the material.

> **Note**
> The answers to all these questions are available at https://github.com/
> PacktPublishing/Accelerate-Model-Training-with-PyTorch-2.X/
> blob/main/quiz/chapter06-answers.md.

Before starting the quiz, remember that it is not a test at all! This section aims to complement your learning process by revising and consolidating the content covered in this chapter.

Choose the correct option for the following questions.

1. What are the two steps to take when simplifying a workflow?

    A. Reduction and compression.

    B. Pruning and reduction.

    C. Pruning and compression.

    D. Reduction and zipping.

2. A pruning technique usually has the following dimensions:

    A. Criterion, scope, and method.

    B. Algorithm, scope, and magnitude.

    C. Criterion, constraints, and targets.

    D. Algorithm, constraints, and targets.

3. Concerning the compression phase, we can assert which of the following?

    A. It receives a compressed model as input and verifies the model's integrity.

    B. It receives a compressed model as input and generates a model partially comprised only of the non-pruned parameters.

    C. It receives a pruned model as input and generates a new brain model comprised only of the non-pruned parameters.

    D. It receives a pruned model as input and evaluates the pruning degree applied to that model.

4. We can execute the model simplifying process on which of the following?

    A. Pre-trained models only.

    B. Pre-trained and non-trained models only.

    C. Non-trained models only.

    D. Non-trained, pre-trained, and trained models.

5. What is one of the main goals of simplifying a trained model?

    A. Accelerate the training process.

    B. Deploy it on resource-constrained environments.

    C. Improve the model's accuracy.

    D. There is no reason to simplify a trained model.

6. Consider the following configuration list passed to the pruner:

    ```
    config_list = [{ 'op_types': ['Conv2d'],
                     'exclude_op_names': ['layer2'],
                     'sparse_ratio': 0.25 }]
    ```

    Which of the following actions would the pruner take?

    A. The pruner will try to nullify 75% of all network parameters.

    B. The pruner will try to nullify 25% of the parameters of all fully connected layers.

    C. The pruner will try to nullify 25% of the parameters of convolutional layers, except the one labeled as "layer2".

    D. The pruner will try to nullify 75% of the parameters of the convolutional layers, except the one labeled as "layer2."

7. What is more likely to happen to the model's accuracy after executing the simplified workflow?

    A. The model's accuracy tends to increase.

    B. The model's accuracy surely increases.

    C. The model's accuracy tends to reduce.

    D. The model's accuracy stays the same.

8. It is necessary to execute a warmup phase before simplifying which of the following?

    A. Non-trained models.

    B. Trained models.

    C. Pre-trained models.

    D. None of the above.

Summary

In this chapter, you learned that simplifying a model by reducing the number of parameters can accelerate the network training process, besides making the model feasible to run on resource-constrained platforms.

Then, we saw that the simplification process consists of two phases: pruning and compression. The former is responsible for determining which parameters must be dropped off from the network, whereas the latter effectively removes the parameters from the model.

Although PyTorch provides an API to prune the model, it is not fully useful to simplify a model. Thus, you were introduced to Microsoft NNI, a powerful toolkit to automate tasks related to deep learning modes. Among the features provided by NNI, this tool offers a complete workflow to simplify a model. All of this is achieved with a couple of new lines added to the original code.

In the next chapter, you will learn how to reduce the numeric precision adopted by the neural network to accelerate the training process and decrease the amount of memory needed to store the model.

7

Adopting Mixed Precision

Scientific computing is a tool that's used by scientists to push the limits of the known. Biology, physics, chemistry, and cosmology are examples of areas that rely on scientific computing to simulate and model the real world. In these fields of knowledge, numeric precision is paramount to yield coherent results. Since each decimal place matters in this case, scientific computing usually adopts double-precision data types to represent numbers with the highest possible precision.

However, that need for extra information comes with a price. The higher the numeric precision, the higher the computing power required to process those numbers. Besides that, higher precision also demands a higher memory space, increasing memory consumption.

In the face of those drawbacks, we must ask ourselves: do we need so much precision to build our models? Usually, we do not! In this sense, we can reduce the numeric precision for a few operations, thus bursting the training process and saving some memory space. Naturally, this process should not affect the model's capacity to make good predictions.

In this chapter, we'll show you how to adopt a mixed precision strategy to burst the model training process without penalizing the model's accuracy. Besides reducing training time in general, this strategy also enables the usage of special hardware resources such as Tensor Cores on NVIDIA's GPU.

Here is what you will learn as part of this chapter:

- The concept of numeric representation in computer systems
- Why lower precision reduces the computational burden of the training process
- How to enable automatic mixed precision on PyTorch

Technical requirements

You can find the complete code for the examples mentioned in this chapter in this book's GitHub repository at https://github.com/PacktPublishing/Accelerate-Model-Training-with-PyTorch-2.X/blob/main.

You can access your favorite environment to execute this code, such as Google Colab or Kaggle.

Remembering numeric precision

Before diving into the benefits of adopting a mixed precision strategy, it is essential to ground you on numeric representation and common data types. Let's start by remembering how computers represent numbers.

How do computers represent numbers?

A computer is a machine – endowed with finite resources – that's designed to work on bits, the smallest unit of information it can manage. As numbers are infinite, computer designers had to put a lot of effort into finding a solution to represent this theoretical concept in a real machine.

To get the work done, computer designers needed to deal with three key factors regarding numeric representation:

- **Sign**: Whether the number is positive or negative
- **Range**: The interval of the represented numbers.
- **Precision**: The number of decimal places.

Considering these elements, computer architects successfully defined numeric data types to represent not only integer and floating-point numbers but also characters, special symbols, and even complex numbers.

Let's consider an example to make things more tangible. Computer architectures and programming languages generally use 32 bits to represent integers via the so-called INT32 format, as shown in *Figure 7.1*:

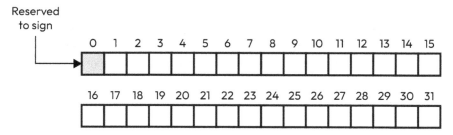

Figure 7.1 – 32-bit numeric representation for integers

Among these 32 bits, 1 bit is reserved to represent the sign of the number, where 0 means positive and 1 means otherwise. The remaining 31 bits are used to represent the number itself. With 31 bits, we can get 2,147,483,648 distinct combinations of zeros and ones. So, the numeric range of this representation falls into -2,147,483,648 and +2,147,483,647. Note that the positive portion has one number fewer because we must represent zero.

This is an example of **signed** integer representation, where 1 bit was used to determine whether the number is positive or negative. However, if only positive numbers are relevant for some cases, it is possible to use an **unsigned** representation instead. The unsigned representation uses all 32 bits to represent the number, resulting in a numeric interval between 0 and 4,294,967,295.

In situations that do not require a larger interval, it is possible to adopt a cheaper format – with only 8 bits – to represent integers: the INT8 representation, as illustrated in *Figure 7.2*. The unsigned version of this representation provides an interval between 0 and 255:

```
0  1  2  3  4  5  6  7
0  0  1  0  1  1  0  0   = 44
```

Figure 7.2 – Example of a number represented in the INT8 format

Assuming that 1 byte is equivalent to 8 bits (this relationship can be different on some computer architectures), the INT32 format demands 4 bytes to represent one integer, whereas INT8 requires only 1 byte to do the same. Therefore, the INT32 format, which is four times more expensive than INT8, demands more memory space and computing power to manipulate those numbers.

The integer representation is quite simple. However, to represent floating-point (decimal) numbers, computer architects had to design a more sophisticated solution, as we will learn in the next section.

Floating-point representation

Modern computers adopt the IEEE 754 standard to represent floating-point numbers. This format defines two types of floating-point representation, namely single and double-precision. **Single-precision**, also known as FP32 or float32, uses 32 bits, while **double-precision**, also known as FP64 or float64, uses 64 bits.

Both single and double-precision are structurally similar and comprise three elements: sign, exponent, and fraction (significand), as shown in *Figure 7.3*.

Figure 7.3 – Structure of floating-point representation

The sign has the same meaning as the integer representation – that is, it defines whether the number is positive or negative. The **exponent** defines the numerical range, and the **fraction** determines the numeric precision – that is, the number of decimal places.

Both formats use 1 bit for the sign. Regarding other portions, FP32 and FP64 use 8 and 11 bits to represent the exponent and 23 and 52 to represent the fraction part, respectively. Roughly speaking, the range of FP64 is slightly higher than FP32 because the former uses 3 bits more than the latter for the exponent part. On the other hand, FP64 provides more than the double precision of FP32 due to 52 bits reserved for the fraction part.

The high numeric precision provided by FP64 makes it suitable for scientific computing, where each additional decimal place is vital to solving the problems tackled in this area. As double precision requires 8 bytes to represent a number, it is commonly used only on tasks that require so much precision. When there is no such requirement, using the single-precision data type is preferable. This is the reason why the training process usually adopts FP32.

Novel data types

Single precision, as defined by the IEEE 754 standard, is the default format for representing floating-point numbers. However, as time passes, the rise of new problems demands new approaches, methods, and data types.

Among the novel data types, we can highlight three that are particularly interesting to machine learning models: FP16, BFP16, and TF32.

FP16

FP16 or float16, as you may have guessed, uses 16 bits to represent floating-point numbers, as depicted in *Figure 7.4*. As it uses half of the 32 bits used on single-precision, this new data type is referred to as **half-precision**.

The structure of FP16 is the same as its siblings FP32 and FP64. The difference lies in the number of bits used to represent the exponent and the fraction parts. FP16 uses 5 and 10 bits to represent the exponent and fraction portions, respectively:

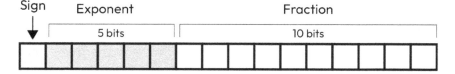

Figure 7.4 – Structure of the FP16 format

FP16 is an alternative to FP32 in cases where the precision provided by float32 is beyond the necessary. In these cases, it is much better to use a simpler data type to save memory space and reduce the computing power needed to manipulate the data.

BFP16

BFP16 or bfloat16 is a novel data type created by Google Brain, an artificial intelligence research group at Google. BFP16, like FP16, uses 16 bits to represent floating-point numbers. However, unlike FP16, the focus of BFP16 is preserving the same range as FP32 while reducing drastically the precision. Thus, BFP16 uses 8 bits to represent the exponent – the same amount used on FP32 – but only 7 bits to represent the fraction part, as shown in *Figure 7.5*:

Figure 7.5 – Structure of the BFP16 format

Google created BFP16 to attend machine learning and artificial intelligence workloads, where precision is not a big deal. At the time of writing, bfloat16 is supported by Intel Xeon processors (through the AVX-512 BF16 instruction set), Google TPUs v2 and v3, NVIDIA GPU A100, and other hardware platforms.

Note that being supported in these hardware platforms, there is no guarantee that bfloat16 is supported and implemented by any software. PyTorch, for example, just supports the usage of bfloat16 on CPUs, even though this data type is also supported by NVIDIA GPUs.

TF32

TF32 stands for TensorFloat 32 and, despite the name, is a 19-bit format created by NVIDIA. The TF32 is a mix of the FP32 and FP16 formats since it uses 8 bits for the exponent and 10 bits for the fraction, just like FP32 and FP16, respectively. Therefore, TF32 allies the precision provided by FP16 with the numeric range of FP32. *Figure 7.6* describes the TF32 format pictorially:

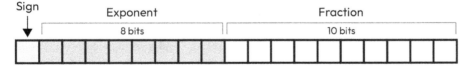

Figure 7.6 – Structure of the TF32 format

Similarly to bfloat16, TF32 was also specifically created to tackle artificial intelligence workloads and is currently supported by newer GPU generations, starting with the Ampere architecture (NVIDIA A100). Besides the benefits of providing a balance between range and precision, TF32 is also supported by Tensor Cores, a special hardware component available on NVIDIA GPUs. We will talk more about Tensor Cores later in this chapter.

A summary, please!

Yes, for sure! That was a lot of information to take in. For this reason, *Table 7.1* summarizes this:

Format	Bits for Exponent	Bits for Fraction	Bytes	Alias
FP32	8	23	4	Float32, single precision
FP64	11	52	8	Float64, double precision
FP16	5	10	2	Float16, half precision
BFP16	8	7	2	Bfloat16
TF32	8	10	4	TensorFloat32

Table 7.1 – Summary of numeric formats

> **Note**
>
> Grigory Sapunov wrote a nice summary about data types. You can find it at https://moocaholic.medium.com/fp64-fp32-fp16-bfloat16-tf32-and-other-members-of-the-zoo-a1ca7897d407.

As you may have noticed, the higher the numeric range and precision, the higher the amount of bytes required to represent the numbers. Thus, the numeric format incurs the amount of resources needed to store and process such numbers.

If we do not need so much precision (and range) to train our models, why not adopt a cheaper format than the usual FP32? If we do that, we will save memory and accelerate the training process as a whole.

Instead of changing the numeric precision of the entire building process, we have the option to adopt a mixed precision approach, as explained next.

Understanding the mixed precision strategy

The benefits of using lower-precision formats are crystal clear. Besides saving memory, the computing power required to handle data with lower precision is less than that needed to process numbers with higher precision.

One approach to accelerate the training process of machine learning models concerns employing a **mixed precision** strategy. Along the lines of *Chapter 6, Simplifying the Model*, we will understand this strategy by asking (and answering, of course) a couple of simple NH questions about this approach.

> **Note**
>
> When searching for information about reducing the precision of deep learning models, you may come across a term known as **model quantization**. Despite being related terms, the goal of mixed precision is quite different from model quantization. The former intends to accelerate the training process by employing reduced numeric precision formats. The latter focuses on reducing the complexity of trained models to use in the inference phase. Thus, be careful to not mistake both terms.

Let's start by answering the most elementary question: what is this strategy about?

What is mixed precision?

As you may have guessed, the mixed precision approach mixes up numeric formats with distinct degrees of precision. This approach aims to try to use a cheaper format **wherever possible** – in other words, it only keeps the default precision where it is mandatory.

In the context of the training process, we seek to mix FP32 – the default numeric format adopted in this task – with a lower-precision representation such as FP16 or BFP16. More specifically, we execute some operations under FP32 and others with a lower-precision format. By doing this, we keep the needed precision on operations where a higher precision is imperative and benefit from the advantages of half-precision representations at the same time.

As illustrated in *Figure 7.7*, the mixed approach is the opposite of the traditional strategy, where we use the same numeric precision for all operations executed in the training process:

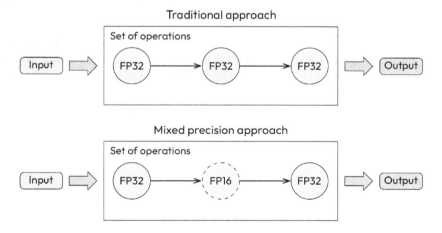

Figure 7.7 – The difference between the traditional and mixed precision approaches

Given the advantages of using a lower-precision format, you might be wondering why not use it on all operations involved in the training process – something like a pure lower-precision approach rather than a mixed-precision strategy, so to speak. It is a fair question, and we will answer it in the following section.

Why use mixed precision?

The question posed here is not about the advantages of using mixed precision, but why we shouldn't use an absolute lower-precision approach.

Well, we cannot use a pure lower-precision approach because of two main reasons:

- Loss of information on the gradient
- Lack of lower-precision operations

Let's take a closer look at each.

Loss of information on the gradient

Reduced precision can lead to **gradient problems**, thus affecting the model's accuracy. As the optimization process is carried, the loss of information on the gradient, derived from reduced precision, can hamper the optimization process as a whole, preventing the model from converging. As a consequence, the trained model can exhibit a lower accuracy.

Shall we clarify this issue? Let's assume we're in a hypothetical situation where we train a model using two distinct precision formats, A and B. Format A supports five decimal places of precision, while format B gives only three decimal places.

Suppose we have trained our model for five training steps. On each training step, the optimizer has calculated the gradient to guide the overall optimization process. However, as shown in *Figure 7.8*, after the third training step, the gradient becomes zero on format B. Thereafter, the optimization process will be blind since the gradient information was lost:

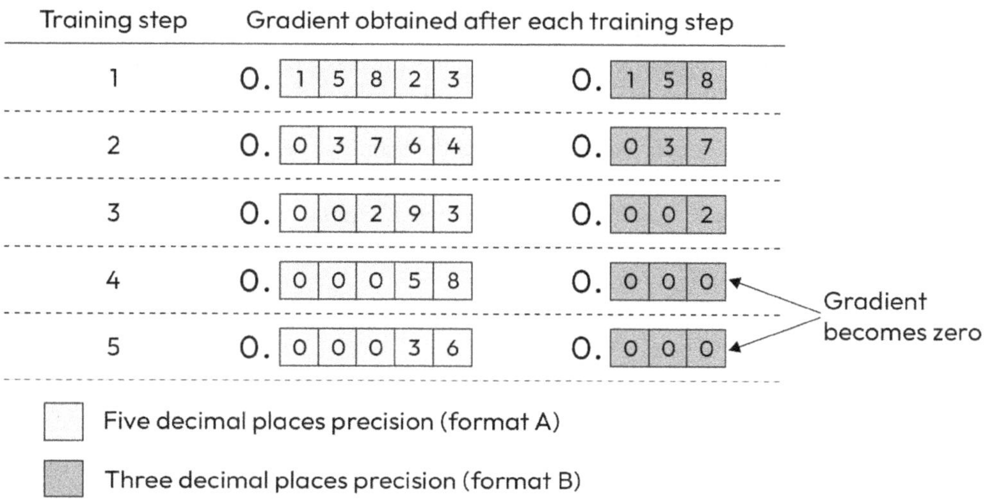

Training step	Gradient obtained after each training step
1	0. [1] [5] [8] [2] [3] 0. [1] [5] [8]
2	0. [0] [3] [7] [6] [4] 0. [0] [3] [7]
3	0. [0] [0] [2] [9] [3] 0. [0] [0] [2]
4	0. [0] [0] [0] [5] [8] 0. [0] [0] [0]
5	0. [0] [0] [0] [3] [6] 0. [0] [0] [0]

Gradient becomes zero

☐ Five decimal places precision (format A)

▨ Three decimal places precision (format B)

Figure 7.8 – Loss of information on the gradient

This is a naive and pictorial example to show what we meant regarding the loss of information on the gradient. Nevertheless, in general terms, this is the problem we can face when opting to use lower-precision formats.

Therefore, we must keep some operations running on the default FP32 format to avoid such problems. Nevertheless, we still need to take care of the gradient when using a lower-precision representation, as we will understand later in this chapter.

Lack of lower-precision operations

Concerning the second reason, we can state that many operations do not have a lower-precision implementation. Besides technical constraints, the cost-benefit ratio of implementing a lower-precision version of some operations can be so low that it is not worth doing it.

For this reason, PyTorch maintains a list of *eligible operations* to run under lower precision to see which operations are currently supported for a given precision and device. For example, the conv2d operation is eligible to run under FP16 on CUDA devices and BFP16 on CPU. On the other hand, the softmax operation is not eligible to run in low-precision, neither on GPU or CPU. Generally speaking, at the time of writing, PyTorch only supports FP16 on CUDA devices and only BFP16 on CPU.

> **Note**
> You can find the complete list of eligible operations to run in lower precision in PyTorch at https://pytorch.org/docs/stable/amp.html.

That said, we must always evaluate whether our model executes at least one of the eligible operations to run with lower precision before jumping headlong into the mixed-precision approach. Otherwise, we will make an unfruitful effort to try to benefit from this strategy.

Even if it is possible to reduce the numeric precision of the training process, we should *not expect a tremendous performance gain* for any scenario. After all, only a subset of operations executed on the training process currently support lower-precision data types. On the other hand, any performance improvement obtained from an effortless process is always welcome.

How to use mixed precision

Usually, we rely on an automatic solution to apply the mixed precision strategy to the training process. This solution is called **automatic mixed precision**, or **AMP** for short.

As illustrated in *Figure 7.9*, AMP automatically evaluates operations that were executed during the training process to decide which ones are eligible to run at lower-precision formats:

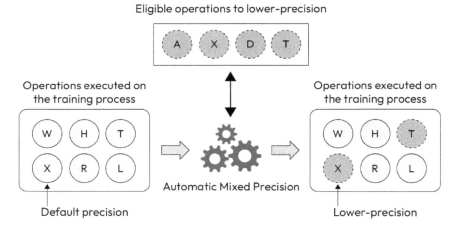

Figure 7.9 – The AMP process

Once AMP finds an operation that matches the requirements to execute at lower precision, it takes the wheel and replaces the operation running at default precision with a lower-precision version. It is an elegant and seamless process that's designed to avoid the occurrence of errors that are difficult to detect, investigate, and fix.

Although it is *strongly recommended to use an automatic solution* to apply the mixed-precision approach, it is possible to do it by hand. However, we must be aware of a few things. Generally speaking, we seek to implement a process manually when the automatic solution does not provide a good or reasonable result, or simply when it does not exist. As the automatic solution is available and there is no guarantee of getting a substantial performance improvement in doing it by hand, we should only consider the manual approach as the last option to adopt mixed precision.

> **Note**
>
> You can always experiment and implement mixed precision manually. It can be a good idea to deepen your knowledge about this topic. You can start by viewing the material presented on NVIDIA GTC 2018, which is available at `https://on-demand.gputechconf.com/gtc-taiwan/2018/pdf/5-1_Internal%20Speaker_Michael%20Carilli_PDF%20For%20Sharing.pdf`.

How about Tensor Cores?

A Tensor Core is a processing unit that's capable of bursting the execution of matrix-to-matrix multiplication, which is an elementary operation that's often executed on AI and HPC workloads. To use this hardware resource, the software (library or framework) must be able to work with the numeric format supported by the Tensor Core. As shown in *Table 7.2*, the numeric format supported by Tensor Core varies according to the GPU architecture:

	Hopper	Ampere	Turing	Volta
Supported Tensor Core precisions	FP64, TF32, bfloat16, FP16, FP8, INT8	FP64, TF32, bfloat16, FP16, INT8, INT4, INT1	FP16, INT8, INT4, INT1	FP16
Supported CUDA Core precisions	FP64, FP32, FP16, bfloat16, INT8	FP64, FP32, FP16, bfloat16, INT8	FP64, FP32, FP16, INT8	FP64, FP32, FP16, INT8

Table 7.2 – Supported data types on Tensor Cores (obtained from NVIDIA's official website)

Tensor Core of newer GPU models, such as Hopper and Ampere (series H and A, respectively), supports lower-precision formats such as TF32, FP16, and bfloat16, and the double-precision format (FP64), which is particularly important to attend traditional HPC workloads.

> **Note**
>
> The Hopper architecture started to support FP8, a fresh new numeric representation that uses only 1 byte to represent floating-point numbers. NVIDIA has created this format to accelerate the training process of Transformer neural networks. The usage of Tensor Cores to run FP8 operations relies on the Transformer Engine library and is outside the scope of this book.

On the other hand, none of the available architectures are equipped with Tensor Cores that support FP32, the default precision format. So, to harness the computing power of this hardware capability, we must adapt our code so that it can use lower-precision formats.

Moreover, the activation of Tensor Cores is conditioned to other factors beyond the adoption of lower-precision representations. Among other things, we must pay attention to the required memory alignment of matrix dimensions for a given combination of architecture, library version, and numeric representation. For example, in the case of A100 (the Ampere architecture), the matrices' dimensions must be multiple of 8 bytes when using FP16 and CuDNN versions before 7.6.3.

Therefore, the adoption of lower precision is the first condition to use Tensor Cores, but it is *not a unique* requirement to properly enable this resource.

> **Note**
>
> You can find more details about the requirements for using Tensor Cores at `https://docs.nvidia.com/deeplearning/performance/dl-performance-matrix-multiplication/index.html#requirements-tc`.

Now that we know about the fundamentals of mixed precision, we can learn how to use this approach in PyTorch.

Enabling AMP

Fortunately, PyTorch provides methods and tools to perform AMP by changing just a few things in our original code.

In PyTorch, AMP relies on enabling a couple of flags, wrapping the training process with the `torch.autocast` object, and using a gradient scaler. The more complex case, which is related to implementing AMP on GPU, takes all these three parts, while the most simple scenario (CPU-based training) requires only the usage of `torch.autocast`.

Let's start by covering the more complex scenario. So, follow me to the next section to learn how to activate this approach in our GPU-based code.

Activating AMP on GPU

To activate AMP on GPU, we need to make three modifications to our code:

1. Enable the CUDA and CuDNN backend flags.
2. Wrap the training loop with `torch.autocast`.
3. Use a gradient scaler.

Let's take a closer look.

Enabling backend flags

As we learned in *Chapter 4, Using Specialized Libraries*, PyTorch relies on third-party libraries (also known as backends in PyTorch's terminology) to help it execute specialized tasks. In the context of AMP, we must enable *four flags* related to CUDA and CuDNN backends. All those flags come disabled by default and should be turned on at the beginning of the code.

> **Note**
> CuDNN is an NVIDIA library that provides optimized operations commonly executed on deep learning neural networks.

The first flag is `torch.backend.cudnn.benchmark`, which activates the benchmark mode of CuDNN. In this mode, CuDNN executes a set of brief tests to determine which operations are the best ones to be executed on a given platform. Although this flag is not directly related to mixed precision, it plays an important role in the process, enhancing the positive effect of AMP.

CuDNN performs this evaluation the first time it is called by PyTorch. In general, this moment occurs at the first training epoch. So, do not be surprised if the first epoch takes more time to execute than the remaining epochs of the training process.

The other two flags are called `cuda.matmul.allow_fp16_reduced_precision_reduction` and `cuda.matmul.allow_bf16_reduced_precision_reduction`. They tell CUDA to use reduced precision on `matmul` operations when executing at FP16 and BFP16 representations, respectively. The `matmul` operation is related to matrix-to-matrix multiplication, which is one of the most essential computing tasks that can be executed on neural networks in general.

The last flag is `torch.backends.cudnn.allow_tf32`, which allows CuDNN to use the TF32 format, thus enabling one of the formats supported by NVIDIA Tensor Cores.

After enabling these flags, we can proceed to change the training loop part.

Wrapping the training loop with torch.autocast

The `torch.autocast` class is responsible for implementing AMP on PyTorch. We can use `torch.autocast` as a context manager or a decorator. Its usage depends on how we have implemented our code. Regardless of the method, AMP works in the same way.

Specifically, in the case of a context manager, we must wrap the forward and loss calculation phases that are executed on the training loop. All other tasks that are performed on the training loop must be left out of the context of `torch.autocast`.

`torch.autocast` accepts four arguments:

- `device_type`: This defines which kind of device the autocast will execute AMP. Accepted values are `cuda`, `cpu`, `xpu`, and `hpu` – that is, the same values we can assign to a `torch.device` object. Naturally, the most common values for this parameter are `cuda` and `cpu`.

- `dtype`: The data type that's used on the AMP strategy. This parameter accepts the corresponding data type object – instantiated from the `torch.dtype` class – related to the data type we want to use on the automatic casting.

- `enabled`: A flag to enable or disable the AMP process. It comes enabled by default, but we can switch it to `false` to debug our code.

- `cache_enabled`: Whether `torch.autocast` should enable the weight cache during the AMP process. This parameter is enabled by default.

The `device_type` and `dtype` parameters are mandatory to use `torch.autocast`. The others are optional and used only for fine-tuning and debugging.

The following excerpt shows the usage of `torch.autocast` as a context manager inside the training loop (for the sake of simplicity, this example shows only the core part of the training loop):

```
with torch.autocast(device_type=device, dtype=torch.float16):
    output = model(images).to(device)
    loss = criterion(output, labels)
```

The other tasks that are executed on the training loop are not encapsulated by `torch.autocast` since we are only interested in applying AMP to the forward and loss calculation phases. Besides that, the remaining tasks of the training process are wrapped by the gradient scaler, as explained next.

Gradient scaler

As we learned at the beginning of this chapter, the usage of lower-precision representations can lead to loss of information on the gradient. To work around this problem, we need to wrap the optimization and backward phases with a **gradient scaler** that's instantiated from `torch.cuda.amp.GradScaler`:

```
scaler = torch.cuda.amp.GradScaler()

with torch.autocast(device_type=device, dtype=torch.float16):
    output = model(images).to(device)
    loss = criterion(output, labels)

optimizer.zero_grad()
scaler.scale(loss).backward()
scaler.step(optimizer)
scaler.update()
```

First, we instantiate an object from `torch.cuda.amp.GradScaler`. Next, we wrap the `optimizer.step()` and `loss.backward()` calls with the gradient scaler so that it assumes control of these tasks. Finally, the training loop asks the scaler to definitively update the parameters of the network.

We'll join these Lego pieces into a unique building block and see what AMP is capable of in the next section!

AMP, show us what you are capable of!

To assess the benefits of using AMP, we will train an EfficientNet neural network architecture, which is available in the `torch.vision.models` package, with the CIFAR-10 dataset.

> **Note**
>
> The complete code shown in this section is available at `https://github.com/PacktPublishing/Accelerate-Model-Training-with-PyTorch-2.X/blob/main/code/chapter07/amp-efficientnet_cifar10.ipynb`.

In this experiment, we will evaluate the usage of AMP with FP16 and BFP16 running under a GPU NVIDIA A100 for 50 epochs. Our baseline execution concerns training EfficientNet under its default precision (FP32), but with the CuDNN benchmark flag enabled. By doing this, we'll make things fair since, despite being important to the AMP process, this flag is not directly related to it.

The baseline execution took 811 seconds to complete, achieving an accuracy equal to 51%. The accuracy itself is not what we care about here; we are interested in evaluating whether AMP will affect the model's quality.

By adopting the BFP16 precision format, the training process took 754 seconds to complete, which represents a shy and disappointing performance improvement of 8%. This happens because eligible operations to run under bfloat16 are implemented only to execute on the CPU, as mentioned previously. Although we are training our model using GPU, some operations still execute under CPU. So, this tiny performance improvement comes from the small pieces of code that continue executing on the CPU.

> **Note**
>
> We're running this experiment with BFP16 on GPU to show the nuances of dealing with AMP. Although PyTorch does not provide BFP16-eligible operations to execute on GPUs, we did not get any warning about doing it. This is an example of how important it is to know the details of the process we are using on our code.

Okay, but how about FP16? Well, the training process running with AMP under Float16 took 486 seconds to complete 50 epochs, representing a *performance gain of 67%*. Due to the work done by the gradient scaler, the accuracy of the model was not affected by the usage of lower-precision formats. Indeed, the model that was trained on this scenario achieved the same 51% accuracy as the baseline code.

We must keep in mind that such performance improvement is just an example of how AMP can accelerate the training process. We can achieve even more impressive results depending on the model, libraries, and devices used in the training process.

The next section provides a couple of questions to help you retain what you have learned in this chapter.

Quiz time!

Let's review what we have learned in this chapter by answering a few questions. Initially, try to answer these questions without consulting the material.

> **Note**
>
> The answers to all these questions are available at `https://github.com/PacktPublishing/Accelerate-Model-Training-with-PyTorch-2.X/blob/main/quiz/chapter07-answers.md`.

Before starting the quiz, remember that this is not a test! This section aims to complement your learning process by revising and consolidating the content covered in this chapter.

Choose the correct option for the following questions:

1. Which of the following numeric formats represents integers by using only 8 bits?

    A. FP8.

    B. INT32.

    C. INT8.

    D. INTFB8.

2. FP16 is a numeric representation that uses 16 bits to represent floating-point numbers. What is this numeric format also known as?

    A. Half-precision floating-point representation.

    B. Single-precision floating-point representation.

    C. Double-precision floating-point representation.

    D. One quarter-precision floating-point representation.

3. Which of the following is a numeric representation for floating-point numbers created by Google to attend to machine learning and artificial intelligence workloads?

    A. GP16.

    B. GFP16.

    C. FP16.

    D. BFP16.

4. NVIDIA created the TF32 data representation. Which of the following number of bits does it use to represent floating-point numbers?

    A. 32 bits.

    B. 19 bits.

    C. 16 bits.

    D. 20 bits.

5. What is the default numeric representation that's used by PyTorch to execute the operations for the training process?

    A. FP32.

    B. FP8.

    C. FP64.

    D. INT32.

6. What is the goal of the mixed precision approach?

    A. Mixed precision tries to adopt lower-precision formats during the training process' execution.

    B. Mixed precision tries to adopt higher-precision formats during the training process' execution.

    C. Mixed precision avoids the usage of lower-precision formats during the training process' execution.

    D. Mixed precision avoids the usage of higher-precision formats during the training process' execution.

7. What are the main advantages of using an AMP approach rather than a manual implementation?

    A. Simple usage and reduction of performance improvement.

    B. Simple usage and reduction of power consumption.

    C. Complex usage and avoidance of errors involving numeric representation.

    D. Simple usage and avoidance of errors involving numeric representation.

8. Besides the lack of lower-precision operations, which of the following options is another reason to not use a pure lower-precision approach in the training process?

    A. Low performance improvement.

    B. High energy consumption.

    C. Loss of information on the gradient.

    D. High usage of main memory.

Summary

In this chapter, you learned that adopting a mixed-precision approach can accelerate the training process of our models.

Although it is possible to implement the mixed precision strategy by hand, it is preferable to rely on the AMP solution provided by PyTorch since it is an elegant and seamless process that's designed to avoid the occurrence of errors involving numeric representation. When this kind of error occurs, they are very hard to identify and solve.

Implementing AMP on PyTorch requires adding a few extra lines to the original code. Essentially, we must wrap the training loop with the AMP engine, enable four flags related to backend libraries, and instantiate a gradient scaler.

Depending on the GPU architecture, library version, and the model itself, we can significantly improve the performance of the training process.

This chapter closes the second part of this book. Next, in the third and last part, we will learn how to spread the training process among multiple GPUs and machines.

Part 3: Going Distributed

In this part, you will learn how to spread the training process across multiple devices and machines. First, you will learn about the fundamental concepts related to the distributed training process. Then, you will learn how to distribute the training process on multiple CPUs in a single machine. After that, you will learn how to train the model by using multiple GPUs in a single machine. At the end, you will learn how to distribute the training process among multiple devices located in multiple machines.

This part has the following chapters:

- *Chapter 8, Distributed Training at a Glance,*
- *Chapter 9, Training with Multiple CPUs*
- *Chapter 10, Training with Multiple GPUs*
- *Chapter 11, Training with Multiple Machines*

8

Distributed Training at a Glance

When we face a complex problem in real life, we usually try to solve it by dividing the big problem into small parts that are easier to treat. So, by combining the partial solutions obtained from the small pieces of the original problem, we reach the final solution. This strategy, called **divide and conquer**, is frequently used to solve computational tasks. We can say that this approach is the basis of the parallel and distributed computing areas.

It turns out that this idea of dividing a big problem into small pieces comes in handy to accelerate the training process of complex models. In cases where using a single resource is not enough to train the model in a reasonable time, the unique way out relies on breaking down the training process and spreading it across multiple resources. In other words, we need to *distribute the training process*.

Here is what you will learn as part of this chapter:

- The basic concepts of distributed training
- The parallel strategies that are used to spread the training process
- The basic workflow to implement distributed training in PyTorch

Technical requirements

You can find the complete code examples mentioned in this chapter in the book's GitHub repository at `https://github.com/PacktPublishing/Accelerate-Model-Training-with-PyTorch-2.X/blob/main`.

You can access your favorite environment to execute the code provided, such as Google Colab or Kaggle.

A first look at distributed training

We'll start this chapter by discussing the reasons for distributing the training process among multiple resources. Then, we'll learn what resources are commonly used to execute this process.

When do we need to distribute the training process?

The most common reason to distribute the training process concerns *accelerating the building process*. Suppose the training process is taking a long time to complete, and we have multiple resources at hand. In that case, we should consider distributing the training process among these various resources to reduce the training time.

The second motivation for going distributed is related to **memory leaks** to load a large model in a single resource. In this situation, we rely on distributed training to allocate different parts of the large model into distinct devices or resources so that the model can be loaded into the system.

However, distributed training is not a *silver bullet* that solves both problems. In many situations, distributed training can achieve the same performance as the traditional execution or, depending on the case, can even be worse. This happens because the overhead that's imposed by preparing the initial setup and performing communication between the multiple resources can overcome the benefits of running the training process in parallel. In addition, we can first try to reduce the model's complexity, as described in *Chapter 6, Simplifying the Model*, instead of moving to the distributed approach. If it succeeds, the resultant model of the simplifying process may now fit on the device.

Therefore, distributed training is not always the correct answer to reduce the training time or to fit the model on a given resource. Thus, it is recommended to take a breath and carefully analyze whether distributed training sounds promising to solve the problem. In short, we can use the flowchart depicted in *Figure 8.1* to decide when to adopt either the traditional or distributed approach:

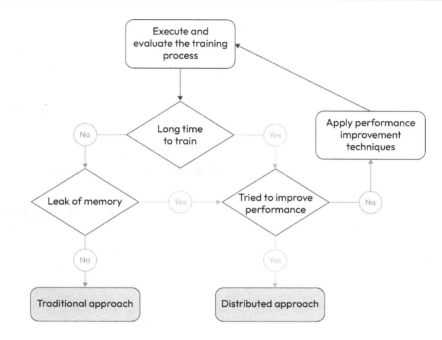

Figure 8.1 – A flowchart for deciding when to use the traditional or distributed approach

In the face of a memory leak or a long time to train, we should apply all possible performance improvement techniques before considering moving to a distributed approach. By doing this, we can avoid problems such as wasting allocated resources that are not being used effectively.

Besides deciding when to distribute the training process, we should evaluate the amount of resources to use in the distributed approach. It is a common mistake to get all available resources to execute the distributed training, supposing that the higher the amount of resources, the shorter the time to train the model. However, there's no guarantee that increasing the amount of resources will result in better performance. The result can be even worse, as we discussed earlier.

In summary, distributed training is useful in cases where the training process takes a long time to finish, or the model does not fit on a given resource. As both cases can be solved by applying performance improvement techniques, we should first try these methods before moving to a distributed strategy. Otherwise, we can face side effects such as resource wastage.

In the next section, we'll provide a higher-level explanation of the computing resources that are used to execute this process.

Where do we execute distributed training?

In a more general way, we can say that distributed training concerns dividing the training process into multiple parts, where each part manages a piece of the entire training process. Each of these parts is allocated to run on a separate computing **resource**.

In the context of distributed training, we can run a part of the training process in the CPU or an accelerator device. Although the accelerator device that's commonly used for this purpose is the GPU, other less popular options, such as FPGA, XPU, and TPU, exist.

These computing resources can be available on a single machine or be located on multiple servers. Moreover, a single machine can possess one or more of these resources.

In other words, we can distribute the training process among *one machine with multiple computing resources* or spread it across *multiple machines with single or multiple resources*. To make this easier to understand, *Figure 8.2* depicts the possible computing resource arrangements you can use in the distributed training process:

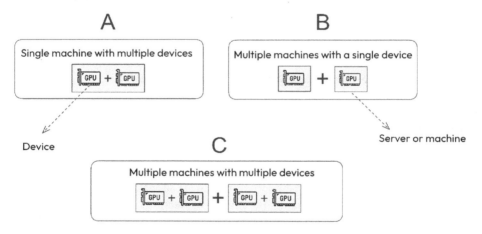

Figure 8.2 – Possible arrangements of computing resources

Arrangement **A**, which has multiple devices located in a single server, is the easiest and fastest configuration to run the distributed training process. As we will learn in *Chapter 11, Training with Multiple Machines*, running the training process on multiple machines depends on the performance delivered by the network used to interconnect the nodes.

Despite the network's performance, the usage of this additional component can downgrade the performance on its own. Therefore, it is preferable to adopt arrangement **A** whenever possible to avoid the use of network interconnection.

Concerning arrangements **B** and **C**, it is better to use the latter because it has a higher ratio of devices per machine. Thus, we can concentrate the distributed training process on a smaller number of machines, hence avoiding network usage.

However, in the absence of arrangements **A** and **C**, it is still a good idea to use arrangement **B**. Even with the bottleneck imposed by the network, it is likely that the distributed training process overcomes the traditional way.

Usually, the GPU is not shared among training instances – that is, the distributed training process allocates one training instance per GPU. In the case of a CPU, things work differently: one CPU can execute more than one training instance. This happens because the CPU is a multicore device, so allocating distinct computing cores to run different training instances is possible.

For example, we can run two training instances in a CPU with 32 computing cores, where each training instance uses half of the available cores, as shown in *Figure 8.3*:

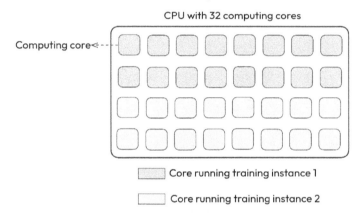

Figure 8.3 – Usage of distinct computing cores to run different training instances

Although it is possible to run distributed training in that way, it is most common to run on multiple GPUs located on a single machine (or more) or in multiple CPUs found on multiple machines. This configuration can be the unique option that's available in many situations, so it is interesting to know how to do it. We will learn more about this in *Chapter 10, Training with Multiple CPUs*.

After being introduced to the distributed training world, it is time to jump to the next section, where you will learn the basic concepts of this approach.

Learning the fundamentals of parallelism strategies

In the previous section, we learned that the distributed training approach divides the whole training process into small parts. As a result, the entire training process can be solved in parallel because each of these small parts is executed simultaneously in distinct computing resources.

The parallelism strategy defines how to divide the training process into small parts. There are two main parallelism strategies: model and data parallelism. The following sections explain both.

Model parallelism

Model parallelism divides the set of operations that are executed during the training process into smaller subsets of computing tasks. By doing this, the distributed process can run these smaller subsets of operations in distinct computing resources, thus accelerating the entire training process.

It turns out that operations executed in the forward and backward phases are not independent of each other. In other words, the execution of one operation usually depends on the output generated by another. Due to this constraint, model parallelism is not straightforward to implement.

Nevertheless, the brilliant human mind has invented three techniques to solve this problem: the **inter-layer**, **intra-operation**, and **inter-operation** paradigms. Let's learn more.

Inter-layer paradigm

In the **inter-layer** paradigm, each model layer is executed in parallel in a distinct computing resource, as shown in *Figure 8.4*:

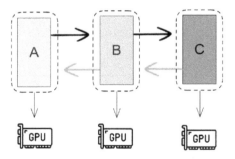

Figure 8.4 – The inter-layer model parallelism paradigm

However, as the computation of a given layer usually depends on the results of another one, the inter-layer paradigm needs to rely on a particular strategy to enable distributed training in these conditions.

When adopting this paradigm, the distributed training process establishes a continuous training flow so that the neural network processes more than one training step at the same time – that is, it processes more than one sample concomitantly. As things go by, the input that's required by one layer at a given training step is already processed in the training flow and is now available to serve as input for that layer.

So, at a given moment, the distributed training process can execute distinct layers in parallel. This process is performed on both the forward and backward phases, which increases the level of parallelism of tasks that can be computed simultaneously even more.

Somehow, this paradigm is very similar to the instruction pipeline technique that's implemented in modern processors, where multiple hardware instructions are executed in parallel. Due to this similarity, the intra-layer paradigm is also called **pipeline parallelism**, where the stages are analogous to the training steps.

Inter-operation paradigm

The **inter-operation** paradigm relies on dividing the set of operations that are executed on each layer into smaller **chunks** of parallelizable computing tasks, as shown in *Figure 8.5*. Each of these chunks of computing tasks is executed on distinct computing resources, hence parallelizing the execution of the layer. After computing all the chunks, the distributed training process combines the partial results obtained from each chunk to yield the layer output:

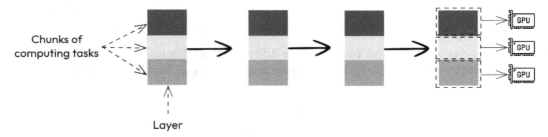

Figure 8.5 – The inter-operation model parallelism paradigm

Because of the dependence between the operations that are executed within the layer, the inter-operation paradigm cannot put dependent operations in distinct chunks. This constraint imposes an additional strain on partitioning the operations into chunks of parallel computing tasks.

For example, consider the graph illustrated in *Figure 8.6*, which represents the computation that's executed in a layer. This graph is comprised of two pieces of input data (rectangles) and four operations (circles), and the arrows indicate the data flow between operations:

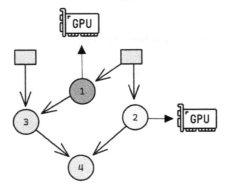

Figure 8.6 – Example of operation partitioning in the inter-operation paradigm

It is easy to see that operations 1 and 2 depend only on the input data, whereas operation 3 needs the output of operation 1 to execute its computation. Operation 4 is the most dependent in the graph since it relies on the results of operations 2 and 3 to be executed.

Therefore, as shown in *Figure 8.6*, the unique partitioning for this graph creates two chunks of parallel operations to run operations 1 and 2 simultaneously. As operations 3 and 4 depend on prior results, they cannot be executed before the completion of other tasks. So, depending on the degree of dependence between operations within the layer, the inter-operation paradigm cannot achieve a higher level of parallelism.

Intra-operation paradigm

The **intra-operation** paradigm splits the execution of an operation into smaller computing tasks, where each computing task applies the operation in a different chunk of input data. In general, the inter-operation approach needs to combine partial results to get the operation done.

While inter-operation runs distinct operations in different computing resources, intra-operation spreads *parts of the same operation* on distinct computing resources, as shown in *Figure 8.7*:

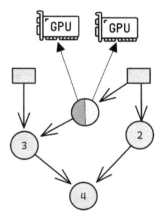

Figure 8.7 – The intra-operation model parallelism paradigm

For example, consider the case in which a layer executes a matrix-to-matrix multiplication, as illustrated in *Figure 8.8*. By adopting the intra-operation paradigm, this multiplication could be divided into two parts, where each part will execute the multiplication on distinct data chunks of matrices A and B. As these partial multiplications are independent of each other, it is feasible to run both tasks concomitantly on different devices:

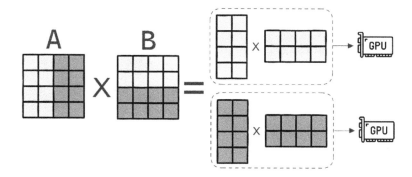

Figure 8.8 – Example of data partitioning in the intra-operation paradigm

After executing these two computing tasks, the intra-operation approach would need to join the partial results to produce the resultant matrix.

Depending on the kind of operation and size of the input data, the intra-operation can achieve a reasonable level of parallelism because more data chunks can be created and submitted to additional computing resources.

However, if the data is too small or the operation is too simple to compute, the overhead of spreading the computation to distinct devices can overcome the potential performance improvement of executing the operation in parallel. This statement is true for both the inter and intra-operation approaches.

Summary

To summarize what we've learned in this section, *Table 8.1* covers the main characteristics of each model parallelism paradigm:

Paradigm	Strategy
Inter-layer	Processes layers in parallel
Intra-operation	Computes distinct operations in parallel
Inter-operation	Computes parts of the same operation in parallel

Table 8.1 – Summary of the model parallelism paradigms

Although model parallelism can accelerate the training process, it has prominent disadvantages, such as poor scalability and imbalance usage of resources, besides being highly dependent on the network architecture. These issues explain why this parallelism strategy is not so popular among data scientists and is usually not the primary option to distribute the training process.

Even so, model parallelism can be the unique solution for cases in which the model does not fit in the computing resource – that is, when we do not have enough memory on the device to allocate the entire model. This is the case with **large language models** (**LLMs**), which commonly have thousands of parameters and occupy a lot of bytes when loaded in memory.

Another strategy, known as data parallelism, is more robust, scalable, and simple to implement, as we will learn in the next section.

Data parallelism

The idea behind the **data parallelism** strategy is very simple to understand. Instead of dividing the set of computing tasks that are executed by the network, the data parallelism strategy divides the training dataset into smaller pieces of data and uses these chunks of data to train distinct *replicas* of the original model, as illustrated in *Figure 8.9*. As each model replica is independent of each other, they can be trained in parallel:

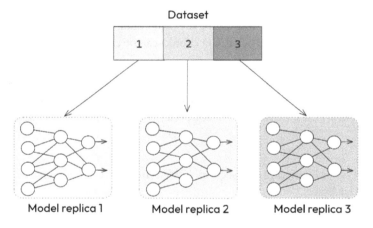

Figure 8.9 – Data parallelism strategy

At the end of each training step, the distributed training process starts a **synchronization phase** to update the weights of all model replicas. This synchronization phase is responsible for collecting and sharing the average gradient among all models running in distinct computing resources. After receiving the average gradient, each replica adjusts its weights according to this shared information.

The synchronization phase is the core mechanism behind the data parallelism strategy. In simpler words, it guarantees that the knowledge obtained by a model replica, after executing a single training step, is shared with the other replicas and vice versa. Thus, when completing the distributed training process, the resultant model has the same knowledge as it would have if trained conventionally:

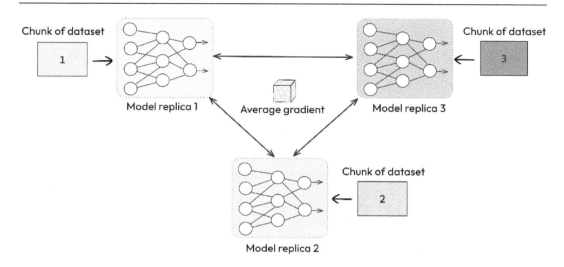

Figure 8.10 – Synchronization phase in data parallelism

There are half a dozen approaches to executing this synchronization phase, including **parameter server** and **all-reduce**. The former does not have good scalability since a unique server is used to aggregate the gradients that are obtained by each model replica, calculate the average gradient, and send it all over the place. As we increase the number of training processes, the parameter server becomes the major bottleneck of the distributed training process.

On the other hand, the all-reduce technique is more scalable because all training instances participate evenly in the update process. Therefore, this technique has been broadly adopted by all frameworks and communications libraries to synchronize the parameters of the distributed training process.

We'll learn more about it in the next section.

All-reduce synchronization

All-reduce is a collective communication technique that's used to simplify the computation that's executed by multiple processes. Since all-reduce is derived from the reduce operation, let's understand this technique before describing the all-reduce communication primitive.

In the context of distributed and parallel computing, the **reduce** operation executes a function on data held in multiple processes and sends the result of this function to a root process. The reduce operation can execute any function, though it is more common to apply trivial and simple functions such as sum, multiplication, average, maximum, and minimum, for example.

Figure 8.11 shows an example of the reduce operation applied to vectors held by four processes. In this example, the reduce primitive executes the sum of the four vectors and sends the result to process 0, which is the root process in this scenario:

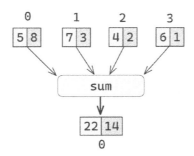

Figure 8.11 – The reduce operation

The all-reduce operation is a particular case of the reduce primitive, in which all processes receive the result of the function, as shown in *Figure 8.12*. So, instead of only sending the result to the root process, all-reduce shares the result with all processes participating in the computation:

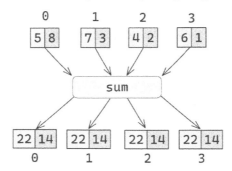

Figure 8.12 – The all-reduce operation

There are different ways to implement the all-reduce operation. Concerning the distributed training context, one of the most efficient solutions is **ring all-reduce**. In this implementation, the processes use a logical ring topology, as shown in *Figure 8.13*, to exchange information among themselves:

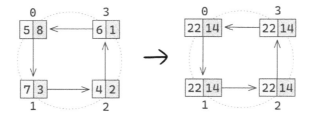

Figure 8.13 – The ring all-reduce implementation

Information flows through the ring until all the processes end up with the same data. There are a couple of libraries that provide an optimized version of the ring all-reduce implementation, such as NCCL from NVIDIA and oneCCL from Intel.

Summary

Data parallelism is easy to understand and implement, besides being flexible and scalable. However, as nothing is perfect, this strategy also has its drawbacks.

Although it provides a higher level of parallelism compared to the model parallelism approach, it can face limiting factors that prevent it from achieving high levels of scalability. As the gradient is shared among all replicas after each training step, any latency in communication between those replicas can slow down the entire training process.

Moreover, the data parallelism strategy does not address the problem of training large models because the model is loaded entirely on the device exactly as-is. The same large model will be loaded with distinct computing resources, which, in turn, will not be able to host them. Concerning models that do not fit on the device, the problem remains the same.

Even so, nowadays, the data parallel strategy is the straightforward approach to distribute the training process. The simplicity and flexibility of this strategy to train a wide range of model types and architectures turns this approach into the default choice to distribute the training process. From now on, we will use the term **distributed training** as a synonym for distributed training based on the data parallelism strategy.

The most adopted frameworks for building machine learning models have a built-in implementation of distributed training. So does PyTorch! In the next section, we will take our first look at how to implement this process.

Distributed training on PyTorch

This section introduces the basic workflow to implement distributed training on PyTorch, besides presenting the components used in this process.

Basic workflow

Generally speaking, the basic workflow to implement distributed training on PyTorch comprises the steps illustrated in *Figure 8.14*:

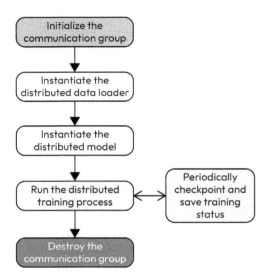

Figure 8.14 – Basic workflow to implement distributed training in PyTorch

Let's look at each step in more detail.

> **Note**
>
> The complete code shown in this section is available at `https://github.com/PacktPublishing/Accelerate-Model-Training-with-PyTorch-2.X/blob/main/code/chapter08/pytorch_ddp.py`.

Initialize and destroy the communication group

The communication group is the logical entity that's used by PyTorch to define and control the distributed environment. So, the first step to code the distributed training concerns *initializing a communication group*. This step is performed by instantiating an object from the `torch.distributed` class and calling the `init_process_group` method, as follows:

```
import torch.distributed as dist
dist.init_process_group()
```

Strictly speaking, the initialization method does not require any argument. However, there are two important parameters, though not mandatory. These parameters allow us to select the communication backend and the initialization method. We will learn about these arguments in *Chapter 9, Training with Multiple CPUs*.

During the creation of the communication group, PyTorch identifies the processes that will participate in the distributed training and assigns a unique identifier to each of them. This identifier, which is called **rank**, can be known by calling the `get_rank` method:

```
my_rank = dist.get_rank()
```

Since *all processes execute the same code*, we can use the rank to differentiate the execution flow of a given process, thus assigning the execution of particular tasks to specific processes. For example, we can use the rank to assign the responsibility of performing the final model evaluation:

```
if my_rank == 0:
    test(ddp_model, test_loader, device)
```

The last step that's executed by distributed training concerns *destroying the communication group*, which was created at the beginning of the code. This process is performed by calling the `destroy_process_group()` method, as follows:

```
dist.destroy_process_group()
```

Terminating the communication group is important since it tells all processes that distributed training is over.

Instantiate the distributed data loader

As we are implementing a data parallelism strategy, it is mandatory to divide the training dataset into small chunks of data to feed each model replica. In other words, we need to instantiate a data loader that's aware of the distributed training process.

In PyTorch, we count on the `DistributedSampler` component to facilitate this task. The `DistributedSampler` component abstracts all unnecessary details from the programmer and is very straightforward to use:

```
from torch.utils.data.distributed import DistributedSampler
dist_loader = DistributedSampler(train_dataset)
train_loader = torch.utils.data.DataLoader(train_dataset,
                                           batch_size=batch_size,
                                           shuffle=False,
                                           sampler=dist_loader)
```

The unique change regards adding an extra parameter, called `sampler`, to the original `DataLoader` creation line. The `sampler` argument must be filled out with an object instantiated from the `DistributedSampler` component, which, in turn, only requires the original dataset object as an input parameter.

The resultant data loader is ready to deal with the distributed training process.

Instantiate the distributed model

With the communication group in hand and the distributed data loader ready to go, it is time to instantiate a distributed version of the original model.

PyTorch provides the native `DistributedDataParallel` component (DDP for short) to encapsulate the original model and prepare it to be trained in a distributed fashion. DDP returns a new model object, which is then used to execute the distributed training process:

```
from torch.nn.parallel import DistributedDataParallel as DDP
model = CNN()
ddp_model = DDP(model)
```

After instantiating the distributed model, all further steps are executed on the distributed version of the model. For example, the optimizer receives the distributed model as a parameter in place of the original model:

```
optimizer = optimizer(ddp_model.parameters(), lr,
                      weight_decay=weight_decay)
```

At this point, we have all we need to run the distributed training process.

Run the distributed training process

Surprisingly, executing the training loop in a distributed manner is almost the same as executing traditional training. The unique difference lies in passing the DDP model as a parameter instead of the original one:

```
train(ddp_model, train_loader, num_epochs, criterion, optimizer,
      device)
```

Nothing else is necessary because the components that we've used so far have intrinsic functionalities to execute the distributed training process.

PyTorch runs the distributed training process continuously until it reaches the defined number of epochs. After completing each training step, PyTorch automatically synchronizes the weights among model replicas. There is no need for any kind of intervention from the programmer's side.

Checkpoint and save the training status

As the distributed training process can take many hours to complete and involves distinct computing resources and devices, it is more likely to be affected by failures.

For this reason, it is recommended to periodically **checkpoint** and **save** the current training state to resume the training process in case of failure. We will cover this topic in detail in *Chapter 10, Training with Multiple GPUs*.

Summary

We may need to instantiate other modules and objects to implement special functionalities for distributed training, but this workflow is usually enough to code a basic – though functional – distributed training implementation.

Communication backend and program launcher

Implementing distributed training on PyTorch involves defining a **communication backend** and using a **program launcher** to execute the process on multiple computing resources.

The following subsections briefly explain each of these components.

Communication backends

As we have learned before, model replicas exchange gradient information among themselves during the distributed training process. From another point of view, the processes running on distinct computing resources must communicate with each other to propagate such data.

In the same way, PyTorch relies on backend software to perform model compiling and multithreading. It also counts on communication backends to provide an optimized communication channel among model replicas.

There are communication backends that specialize in working with high-performance networks, whereas other ones are suitable for dealing with communication among multiple devices inside a single machine.

The most common communication backends that are supported by PyTorch are Gloo, MPI, NCCL, and oneCCL. Each of these backends is particularly interesting to use in a given scenario, as we will learn in the next few chapters.

Program launchers

Running distributed training is not the same as executing a traditional training process. The execution of any distributed and parallel program is quite distinct from running any traditional and sequential program.

In the context of distributed training in PyTorch, we use program launchers to put the distributed process on the road. This tool is responsible for setting up the environment and creating processes in the operating system, local or remote.

The most common launchers that are used for this are **torchrun** and **mpirun**, though it is possible to use older methods such as mp.spawn, which is provided by the torch.multiprocessing package.

Putting everything together

The concept map illustrated in *Figure 8.15* shows the components and resources that surround the distributed training process provided by PyTorch:

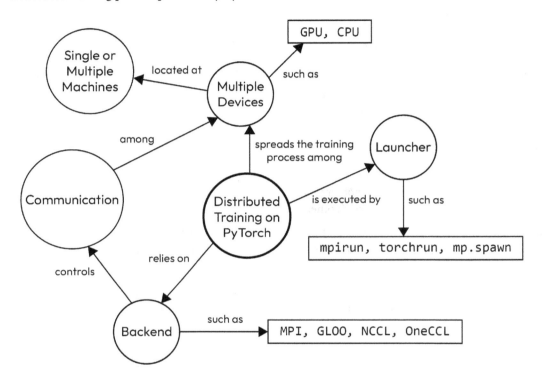

Figure 8.15 – Concept map of the distributed training in PyTorch

As we have learned, PyTorch relies on a communication backend to control communication among multiple computing resources and uses a program launcher to submit distributed training to the local or remote operating system.

There are distinct ways to do the same thing. For example, we can use a certain program launcher to execute distributed training based on two different communication backends. The contrary is also true – that is, there are cases in which a communication backend supports more than one launcher.

So, defining the tuple *communication backend x program launcher* will depend on the environment and resources used in the distributed training process. We will learn more about this in the next few chapters.

The next section provides a couple of questions to help you retain what you have learned in this chapter.

Quiz time!

Let's review what we have learned in this chapter by answering a few questions. Initially, try to answer these questions without consulting the material.

> **Note**
>
> The answers to all these questions are available at `https://github.com/PacktPublishing/Accelerate-Model-Training-with-PyTorch-2.X/blob/main/quiz/chapter08-answers.md`.

Before starting the quiz, remember that this is not a test! This section aims to complement your learning process by revising and consolidating the content covered in this chapter.

Choose the correct option for the following questions.

1. What are the two main reasons for distributing the training process?

   A. Reliability and performance improvement.

   B. Leak of memory and power consumption.

   C. Power consumption and performance improvement.

   D. Leak of memory and performance improvement.

2. Which are the two main parallel strategies to distribute the training process?

   A. Model and data parallelism.

   B. Model and hardware parallelism.

   C. Hardware and data parallelism.

   D. Software and hardware parallelism.

3. Which paradigm is used by the model parallelism approach?

   A. Inter-model.

   B. Inter-data.

   C. Inter-operation.

   D. Inter-parameter.

4. What does the intra-operation paradigm process in parallel?

   A. Distinct operations.

   B. Parts of the same operation.

    C.   Layers of the model.

    D.   Dataset samples.

5.   Besides the parameter server, what other synchronization approach is used by the data parallelism strategy?

    A.   All-operations.

    B.   All-gather.

    C.   All-reduce.

    D.   All-scatter.

6.   What is the first step of executing distributed training in PyTorch?

    A.   Initialize the communication group.

    B.   Initialize the model replica.

    C.   Initialize the data loader.

    D.   Initialize the container environment.

7.   In the context of distributed training in PyTorch, which component is used to put the distributed process on the road?

    A.   Execution library.

    B.   Communication backend.

    C.   Program launcher.

    D.   Compiler backend.

8.   PyTorch supports which of the following as a communication backend?

    A.   NDL.

    B.   MPI.

    C.   AMP.

    D.   NNI.

Summary

In this chapter, you learned that distributed training is indicated to accelerate the training process and training models that do not fit on a device's memory. Although going distributed can be a way out for both cases, we must consider applying performance improvement techniques before going distributed.

We can perform distributed training by adopting the model parallelism or data parallelism strategy. The former employs different paradigms to divide the model computation among multiple computing resources, while the latter creates model replicas to be trained over chunks of the training dataset.

We also learned that PyTorch relies on third-party components such as communication backends and program launchers to execute the distributed training process.

In the next chapter, we will learn how to spread out the distributed training process so that it can run on multiple CPUs located in a single machine.

9

Training with Multiple CPUs

When accelerating the model-building process, we immediately think of machines endowed with GPU devices. What if I told you that running distributed training on machines equipped only with multicore processors is possible and advantageous?

Although the performance improvement obtained from GPUs is incomparable, we should not disdain the computing power provided by modern CPUs. Processor vendors have continuously increased the number of computing cores on CPUs, besides creating sophisticated mechanisms to treat access contention to shared resources.

Using CPUs to run distributed training is especially interesting for cases where we do not have easy access to GPU devices. Thus, learning this topic is vital to enrich our knowledge about distributed training.

In this chapter, we show how to execute the distributed training process on multiple CPUs in a single machine by adopting a general approach and using the Intel oneCCL backend.

Here is what you will learn as part of this chapter:

- The advantages of distributing training on multiple CPUs
- How to distribute the training process among multiple CPUs
- How to burst the distributed training by using Intel oneCCL

Technical requirements

You can find the complete code of examples mentioned in this chapter in the book's GitHub repository at https://github.com/PacktPublishing/Accelerate-Model-Training-with-PyTorch-2.X/blob/main.

You can access your favorite environments to execute this notebook, such as Google Colab or Kaggle.

Why distribute the training on multiple CPUs?

At first sight, thinking about distributing the training process among multiple CPUs in a single machine sounds slightly confusing. After all, we could increase the number of threads used in the training process to allocate all available CPUs (computing cores).

However, as said by Carlos Drummond de Andrade, a famous Brazilian poet, "*In the middle of the road there was a stone. There was a stone in the middle of the road.*" Let's see what happens to the training process when we just increase the number of threads in a machine with multiple cores.

Why not increase the number of threads?

In *Chapter 4, Using Specialized Libraries*, we learned that PyTorch relies on OpenMP to accelerate the training process by employing the multithreading technique. OpenMP assigns threads to physical cores intending to improve the performance of the training process.

So, if we have a certain number of available computing cores, why not increase the number of threads used in the training process rather than thinking about going distributed? The answer is quite simple, actually.

PyTorch has a *limit on the level of parallelism* it can achieve in running the training processes when using multithreads. This limit means there will not be a performance improvement after crossing a certain number of threads. In simpler terms, after a certain threshold, the training time will be the same, no matter how many extra cores we use to train the model.

This behavior is not exclusive to the training process executed by PyTorch. It is very common in many kinds of parallel applications. Depending on the problem and the design of the parallel strategy, increasing the number of threads can turn the parallel task into being so small and simple to execute that the benefits of parallelizing the problem will be suppressed by the overhead of controlling the execution of each parallel task.

Let's see a practical example of this behavior. *Table 9.1* presents the execution time of training a CNN model against the CIFAR-10 dataset over five epochs using a machine equipped with 16 physical cores:

Threads	Execution Time
1	311
2	189
4	119
8	93
12	73
16	73

Table 9.1 – Execution time of training process

As shown in *Table 9.1*, there is no difference in the execution time whether using 12 or 16 cores to train the model. Due to the limit imposed by the parallelism level, PyTorch is stuck on the same execution time despite increasing the number of cores by more than 30%. Moreover, even when the training process used 50% more threads (8 to 12), the performance improvement was less than 27%.

These results pinpoint that using more than eight threads to execute the training process will not significantly reduce the execution time in this case. Consequently, we will incur resource wastage because PyTorch allocates a given number of cores that do not contribute to accelerating the training process. Actually, a higher number of threads can slow down the training process since it can increase the overhead imposed by communication and control tasks.

To work around this opposite effect, we should consider distributing the training process by running distinct training instances on the same machine. Instead of looking at the code, let's jump directly to the results so you can see the benefits of this strategy!

Distributed training on rescue

We conducted the following tests by using the same model, parameters, and dataset as the previous experiment. Naturally, we used the same machine as well.

In the first test, we created two instances of the distributed training process, each using eight cores, as shown in *Figure 9.1*:

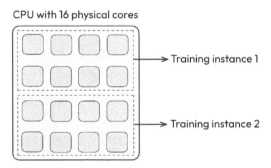

Figure 9.1 – Allocation of distributed training instances

The distributed training process took 58 seconds to complete, representing an *improvement of 26%* in the time needed to execute the model-building process. We have reduced the execution time by more than 25% by adopting the parallel data strategy technique. Nothing else had changed, neither in the hardware capacity nor in the software stack. Furthermore, the performance improvement can be even higher for machines with more computing cores.

However, as we have been saying throughout the book, everything usually has a cost. In this case, the cost is related to the model accuracy. The traditional training process built a model with an accuracy equal to 45.34%, whereas the model created by the distributed training achieved an accuracy of 44.01%.

Although the difference is tiny (around 1.33%), we should not ignore it because there is a relation between model accuracy and the number of distributed training instances.

Table 9.2 shows the results of tests involving different combinations of training instances (processes of the distributed training) and the number of threads used by each training instance. As the tests were executed in a machine with 16 physical cores, and when considering numbers to the power of 2, we have three possible combinations of training instances and threads:

Training instances	Number of threads	Execution time	Accuracy
2	8	58	44.01%
4	4	45	40.11%
8	2	37	38.63%

Table 9.2 – Execution time of distributed training process

As we can verify from *Table 9.2*, the higher the number of training instances, the lower the model accuracy. This behavior is expected because model replicas update their parameters accordingly to an *average* gradient, which results in a loss of information concerning the optimization process.

Conversely, the execution time decreases as the number of training instances increases. When running eight training instances with two threads each, the distributed training process took only 37 seconds to complete, which is almost *two times faster* than running the traditional training with 16 threads. As a counterpart, the accuracy decreased from 45% to 39%.

Undeniably, distributing the training process among multiple processing cores is advantageous in terms of accelerating the training process. We should only take care of model accuracy.

In the next section, we will learn how to code and run the distributed training on multiple CPUs.

Implementing distributed training on multiple CPUs

This section shows how to implement and run the distributed training on multiple CPUs using **Gloo**, a simple yet powerful communication backend.

The Gloo communication backend

In *Chapter 8*, *Distributed Training at a Glance*, we learned PyTorch relies on backends to control the communication among devices and machines involved in distributed training.

The most basic communication backend supported by PyTorch is called Gloo. This backend comes with PyTorch by default and does not require any particular configuration. The Gloo backend is a collective communication library created by Facebook, and it is now an open-source project governed by the BSD license.

> **Note**
>
> You can find the source code of Gloo at `http://github.com/facebookincubator/gloo`.

As Gloo is very simple to use and is available by default on PyTorch, it appears to be the first option to run the distributed training in an environment comprising only CPUs and machines interconnected by a regular network. Let's see this backend in practice in the following sections.

Coding distributed training to run on multiple CPUs

This section presents the code to run the distributed training process on a *single machine with multiple computing cores*. The code is pretty much the same as the one presented in *Chapter 8, Distributed Training at a Glance*, except for a few particularities related to the context of this scenario.

> **Note**
>
> The complete code shown in this section is available at `https://github.com/PacktPublishing/Accelerate-Model-Training-with-PyTorch-2.X/blob/main/code/chapter09/gloo_distributed-cnn_cifar10.py`.

For this reason, this section describes the main changes required to adjust the basic workflow described in *Chapter 8, Distributed Training at a Glance*, to make it feasible to run the distributed training on multiple cores. Essentially, we need to perform two modifications, as explained in the following two sections.

Initialization of the communication group

The *first modification* is related to the initialization of the communication group. Instead of calling `dist.init_process_group` without parameters, this implementation will pass two arguments, as we have mentioned in *Chapter 8, Distributed Training at a Glance*:

```
dist.init_process_group(backend="gloo", init_method="env://")
```

The `backend` argument tells PyTorch which communication backend it must use to control the communication among multiple training instances. In this primary example, we will use Gloo as the communication backend. So, we just need to pass a lowercase string of the backend's name to the parameter.

> **Note**
>
> To check whether a backend is available, we can execute the `torch.distributed.is_<backend>_available()` command. For example, to verify if Gloo is available on the current PyTorch environment, we just need to call `torch.distributed.is_gloo_available()`. This method returns `True` when it is available and `False` if not.

The second parameter, named `init_method`, defines the initialization method used by PyTorch to create the communication group. The method tells PyTorch how to get the information it needs to initialize the distributed environment.

Nowadays, there are three possible methods to inform the configuration required to initialize the communication group:

- **TCP**: Use a specified IP address and TCP port
- **Shared file system**: Use a file system that is accessible to all processes participating in the communication group
- **Environment variables**: Use the environment variables defined on the scope of the operating system

As you might guess, the `env://` value, which is used in this example, refers to the third method to initialize the communication group, i.e., the environment variables option. In the next section, we will learn which environment variables we use to set up the communication group. For now, it is essential to remember how PyTorch gets the information it needs to establish the communication group.

CPU allocation map

The *second modification* refers to defining the allocation of threads belonging to each training instance to different cores. By doing this, we guarantee that all threads use exclusive computing resources and do not compete for a given processing core.

To explain what this means, let's use a practical example. Suppose we want to run the distributed training in a machine with 16 physical cores. We decided to run two training instances, each one using eight threads. If we do not take care of the allocation of these threads, both training instances may compete for a given computing core, leading to a performance bottleneck. This is precisely the opposite of what we want.

To avoid this problem, we must define the allocation map for all threads at the beginning of the code. The following snippet of code shows how to do it:

```
import os
num_threads = 8
index = int(os.environ['RANK']) * num_threads
cpu_affinity = "{}-{}".format(index, (index + num_threads) - 1)
os.environ['OMP_NUM_THREADS'] = "{}".format(num_threads)
```

```
os.environ['KMP_AFFINITY'] = \
    "granularity=fine,explicit,proclist=[{}]".format(cpu_affinity)
```

> **Note**
>
> It is essential to remember that all communication group processes execute the same code. If we need to define a different execution flow for the processes, we must use the rank.

Let's take this line by line to understand what this code is doing.

We start by defining the number of threads used by each training instance, i.e., by each process participating in the distributed training:

```
num_threads = 8
```

Next, we calculate the index of the process considering its rank and the number of threads. The rank is obtained from an environment variable called RANK, which is properly defined by the program launcher:

```
index = int(os.environ['RANK']) * num_threads
```

This index is used to identify the first processing core allocated to that process. For example, when considering the case of 8 threads and two processes, the processes identified by ranks 0 and 1 will have indexes equal to 0 and 8, respectively.

Starting from that index, each process will allocate the subsequent cores to its threads. So, by taking the previous scenario as an example, the process with rank 0 will assign its threads to computing cores 0, 1, 2, 3, 4, 5, 6, and 7. Likewise, the process with rank 1 will use the computing cores 8, 9, 10, 11, 12, 13, 14, and 15.

As OpenMP accepts an interval list format as input for setting the CPU affinity, we can define the allocation map by indicating the first and last cores. The first core is the index, and the last core is obtained by summing up the index with the number of threads and then subtracting from 1:

```
cpu_affinity = "{}-{}".format(index, (index + num_threads) - 1)
```

When considering our example, the process with rank 0 and 1 will set the variable cpu_affinity with values "0-7" and "8-15", respectively.

The last two lines of our piece of code define the OMP_NUM_THREADS and KMP_AFFINITY environment variables according to the values we obtained before:

```
os.environ['OMP_NUM_THREADS'] = "{}".format(num_threads)
os.environ['KMP_AFFINITY'] = \
    "granularity=fine,explicit,proclist=[{}]".format(cpu_affinity)
```

As you should remember, those variables are used to control the behavior of OpenMP. The OMP_NUM_ THREADS variable tells OpenMP the number of threads to use in multithreading, and KMP_AFFINITY defines the CPU affinity for these threads.

These two modifications are enough to adjust the basic workflow presented in *Chapter 8, Distributed Training at a Glance*, to execute the distributed training on multiple CPUs.

With the code ready to execute, the subsequent step concerns defining the program launcher and configuring the parameters to launch the distributed training.

Launching distributed training on multiple CPUs

As we have learned in *Chapter 8, Distributed Training at a Glance*, PyTorch relies on a program launcher to set up the distributed environment and create the processes needed to run the distributed training.

For this scenario, we will use torchrun, which is a native PyTorch launcher. Besides being simple to use, torchrun is already available on the default PyTorch installation. Let's look at more details about this tool.

torchrun

Roughly speaking, torchrun performs two main tasks: it *defines the environment variables related to the distributed environment* and *instantiates the processes on the operating system*.

torchrun defines a set of environment variables to inform PyTorch about the parameters it needs to initialize the communication group. After setting the appropriate environment variables, torchrun creates the processes that will participate in the distributed training.

> **Note**
> Besides these two main tasks, torchrun provides more advanced functionalities such as resuming a failed training process or dynamically adjusting the resources used in the training phase.

To run the distributed training in a single machine, torchrun requires a few parameters:

- nnodes: number of nodes used in the distributed training
- nproc-per-node: number of processes to run in each machine
- master-addr: IP address of the machine used to run the distributed training

The command to execute torchrun for our example is the following:

```
maicon@packt:~$ torchrun --nnodes 1 --nproc-per-node 2 --master-addr
localhost pytorch_ddp.py
```

As the distributed training will run in a single machine, we set the nnodes parameter to 1 and the master-addr argument to localhost, which is the alias name for the local machine. In this example, we desire to run two training instances; hence, the parameter nproc-per-node is set to 2.

From these parameters, torchrun will set the appropriate environment variables and instantiate two processes on the local operating system to run the program pytorch_ddp.py, as shown in *Figure 9.2*:

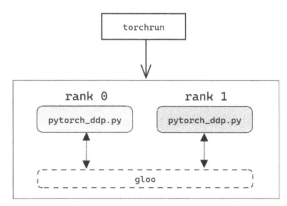

Figure 9.2 – Scheme of torchrun execution

As illustrated in *Figure 9.2*, each process has its rank and communicates to each other through Gloo. Moreover, each process will create eight threads, and each thread will run on a distinct physical core, as defined in the CPU allocation map. These processes will act as different instances of the distributed training process despite being executed in the same machine and running upon the same CPU die.

To make things easier, we can create a bash script to facilitate the usage of torchrun in different situations. Let's learn how to do it in the next section.

Launching script

Instead of typing the torchrun command several times for different scenarios, we can create a bash script to simplify the launch of the distributed training process and run it on a single machine with multiple computing cores.

An example of this launching script is shown here:

```
TRAINING_SCRIPT=$1
NPROC_PER_NODE=$2
NNODES= "1"
MASTER_ADDR= "localhost"
TORCHRUN_COMMAND="torchrun --nnodes $NNODES --nproc-per-node $NPROC_
PER_NODE --master-addr $MASTER_ADDR $TRAINING_SCRIPT"
$TORCHRUN_COMMAND
```

This script sets immutable parameters, such as nnodes and master-addr, with default values and leaves the customizable arguments, such as the name of the program and nproc-per-node, open to be defined in the execution line. So, to run our previous example, we just need to execute the following command:

```
maicon@packt:~$ ./launch_multiple_cpu.sh pytorch_ddp.py 2
```

The launch_multiple_cpu.sh script will call torchrun with the appropriate set of parameters. As you might imagine, it is effortless to change the arguments of this script to use it with another training program, as well as to run a different number of training instances.

Furthermore, we can adapt this script to use it along with container images provided by solutions such as Apptainer and Docker. So, instead of calling torchrun directly on the command line, the script could be modified to execute torchrun inside a container image:

```
TRAINING_SCRIPT=$1
NPROC_PER_NODE=$2
SIF_IMAGE=$3
NNODES= "1"
MASTER_ADDR= "localhost"
TORCHRUN_COMMAND="torchrun --nnodes $NNODES --nproc-per-node $NPROC_
PER_NODE --master-addr $MASTER_ADDR $TRAINING_SCRIPT"
apptainer exec $SIF_IMAGE $TORCHRUN_COMMAND
```

By considering a container image named pytorch.sif, the command line of this new version of local_launch will be the following:

```
maicon@packt:~$ ./launch_multiple_cpu_container.sh pytorch_ddp.py 2
pytorch.sif
```

In the next section, we will learn how to run this same distributed training process but using Intel oneCCL as the communication backend.

Getting faster with Intel oneCCL

The results shown in *Table 9.2* attest that Gloo fulfills the role of the communication backend for the distributed training process in PyTorch very well.

Even so, there is another option for the communication backend to go even faster on Intel platforms: the Intel oneCCL collective communication library. In this section, we will learn what this library is and how to use it as a communication backend for PyTorch.

What is Intel oneCCL?

Intel oneCCL is a collective communication library created and maintained by Intel. Along the lines of Gloo, oneCCL also provides collective communication primitives such as the so-called "All-reduce."

Naturally, Intel oneCCL is optimized to run on Intel platform environments, though this does not necessarily mean it will not work on other platforms. We can use this library to provide collective communication among the processes executing in the same machine (intraprocess communication) or the processes running in multiple-node (interprocess communication.

Although its primary usage lies in providing collective communication for deep learning frameworks and applications, oneCCL can also be used by any distributed program written in C++ or Python.

Like Intel OpenMP, the Intel oneCCL does not come by default with the regular PyTorch installation; we need to install it on our own. When considering a pip-based environment, we can easily install oneCCL by executing the following command:

```
pip install oneccl_bind_pt==2.1.0 --extra-index-url https://pytorch-extension.intel.com/release-whl/stable/cpu/us/
```

After installing oneCCL, we are ready to incorporate it into our code and launch the distributed training. Let's see how to do this in the subsequent sections.

> **Note**
>
> You can find more information about Intel oneCCL at https://oneapi-src.github.io/oneCCL/.

Code implementation and launching

To use Intel oneCCL as a communication backend, we must change a few parts of the code presented in the previous section.

> **Note**
>
> The complete code shown in this section is available at https://github.com/PacktPublishing/Accelerate-Model-Training-with-PyTorch-2.X/blob/main/code/chapter09/oneccl_distributed-cnn_cifar10.py.

The first modification concerns importing an artifact and setting three environment variables:

```
import oneccl_bindings_for_pytorch
os.environ['CCL_PROCESS_LAUNCHER'] = "torch"
os.environ['CCL_ATL_SHM'] = "1"
os.environ['CCL_ATL_TRANSPORT'] = "ofi"
```

These environment variables configure the behavior of oneCCL. The CCL_PROCESS_LAUNCHER parameter talks to oneCCL, which launches it. In our case, we must set this environment variable to torch since PyTorch is calling oneCCL. Environment variables CCL_ATL_SHM and CCL_ATL_TRANSPORT, when set to 1 and ofi, respectively, configure oneCCL to use the shared memory as the means to provide communication among processes.

Shared memory is an interprocess communication technique.

> **Note**
> You can dive into the environment variables of Intel oneCCL by consulting this website: https://oneapi-src.github.io/oneCCL/env-variables.html.

The second modification is related to changing the backend set in the initialization of the communication group:

```
dist.init_process_group(backend="ccl", init_method="env://")
```

The rest of the code and the launching method are equal to Gloo's code. We can set the CCL_LOG_LEVEL to debug or trace environment variable to verify whether oneCCL is being used.

After making those modifications, you may wonder if oneCCL is worth it. Let's find out in the next section.

Is oneCCL really better?

As shown in *Table 9.3*, oneCCL has accelerated our training process by approximately 10% when compared to Gloo's implementation. If compared to the traditional execution with 16 threads, the performance improvement with oneCCL achieved almost 40%:

		oneCCL		Gloo	
Training instances	Number of threads	Execution time	Accuracy	Execution time	Accuracy
2	8	53	43.12%	58	44.01%
4	4	42	41.03%	45	40.11%
8	2	35	37.99%	37	38.63%

Table 9.3 – Execution time of distributed training process running under Intel oneCCL and Gloo

Regarding the model's accuracy, the distributed training with oneCCL and Gloo practically achieved the same results for all scenarios.

So, the question that comes to our mind is, When do we use one backend or another? If we are using an Intel-based environment, then oneCCL is preferable. After all, the training process with Intel oneCCL was 10 % faster than using Gloo.

On the other hand, Gloo comes by default with PyTorch, is very simple to use, and achieves a reasonable performance improvement. So, if we are not training in an Intel platform nor seeking the maximum possible performance, Gloo is a good choice.

The next section provides a couple of questions to help you retain what you have learned in this chapter.

Quiz time!

Let's review what we have learned in this chapter by answering a few questions. At first, try to answer these questions without consulting the material.

> **Note**
>
> The answers to all these questions are available at `https://github.com/PacktPublishing/Accelerate-Model-Training-with-PyTorch-2.X/blob/main/quiz/chapter09-answers.md`.

Before starting the quiz, remember that it is not a test at all! This section aims to complement your learning process by revising and consolidating the content covered in this chapter.

Choose the correct option for the following questions.

1. In multicore systems, we can improve the performance of the training process by increasing the number of threads used by PyTorch. Concerning this topic, we can affirm which of the following?

    A. After crossing a certain number of threads, the performance improvement can deteriorate or stay the same.

    B. The performance improvement always keeps rising, no matter the number of threads.

    C. There is no performance improvement when increasing the number of threads.

    D. Performance improvement is only achieved when using 16 threads.

2. Which is the most basic communication backend supported by PyTorch?

    A. NNI.

    B. Gloo.

C. MPI.

D. TorchInductor.

3. Which is the default program launcher provided by PyTorch?

A. PyTorchrun.

B. Gloorun.

C. MPIRun.

D. Torchrun.

4. In the context of PyTorch, what is Intel oneCCL?

A. Communication backend.

B. Program launcher.

C. Checkpointing automation tool.

D. Profiling tool.

5. When considering a non-Intel environment, what would be the most reasonable choice for the communication backend?

A. Gloorun.

B. Torchrun.

C. oneCCL.

D. Gloo.

6. Concerning the performance of the training process when using Gloo or oneCCL as a communication backend, we can say which of the following?

A. There is no difference at all.

B. Gloo is always better than oneCCL.

C. oneCCL can overcome Gloo in Intel platforms.

D. oneCCL is always better than Gloo.

7. When distributing the training process among multiple CPUs and cores, we need to define the allocation of threads in order to do which of the following?

A. Guarantee all threads have exclusive usage of computing resources.

B. Guarantee secure execution.

C. Guarantee protected execution.

D. Guarantee that data are shared among all threads.

8. What are the two main tasks of torchrun?

A. Create a pool of shared memory and instantiate the processes in the operating system.

B. Define the environment variables related to the distributed environment and instantiate the processes on the operating system.

C. Define the environment variables related to the distributed environment and create a pool of shared memory.

D. Identify the best number of threads to run with PyTorch.

Summary

In this chapter, we learned that distributing the training process on multiple computing cores can be more advantageous than increasing the number of threads used in traditional training. This happens because PyTorch can face a limit on the parallelism level employed in the regular training process.

To distribute the training among multiple computing cores located in a single machine, we can use Gloo, a simple communication backend that comes by default with PyTorch. The results showed that the distributed training with Gloo achieved a performance improvement of 25% while retaining the same model accuracy.

We also learned that oneCCL, an Intel collective communication library, can accelerate the training process even more when executed on Intel platforms. With Intel oneCCL as the communication backend, we reduced the training time by more than 40%. If we are willing to reduce the model accuracy a little bit, it is possible to train the model two times faster.

In the next chapter, we will learn how to spread out the distributed training process to run on multiple GPUs located in a single machine.

10
Training with Multiple GPUs

Undoubtedly, the computing power provided by GPUs is one of the factors that's responsible for boosting the deep learning area. If a single GPU device can accelerate the training process exceedingly, imagine what we can do with a multi-GPU environment.

In this chapter, we will show you how to use multiple GPUs to accelerate the training process. Before describing the code and launching procedure, we will dive into the characteristics and nuances of the multi-GPU environment.

Here is what you will learn as part of this chapter:

- The fundamentals of a multi-GPU environment
- How to distribute the training process among multiple GPUs
- NCCL, the default backend for distributed training on NVIDIA GPUs

Technical requirements

You can find the complete code mentioned in this chapter in this book's GitHub repository at `https://github.com/PacktPublishing/Accelerate-Model-Training-with-PyTorch-2.X/blob/main`.

You can access your favorite environment to execute this code, such as Google Colab or Kaggle.

Demystifying the multi-GPU environment

A multi-GPU environment is a computing system with more than one GPU device. Although multiple interconnected machines with just one GPU can be considered a multi-GPU environment, we usually employ this term to describe environments with two or more GPUs per machine.

To understand how this environment works under the hood, we need to learn about the connectivity of the devices and technologies that are adopted to provide efficient communication across multiple GPUs.

However, before we dive into these topics, we will answer a disquieting question that has probably come to your mind: will we have access to an expensive environment like that? Yes, we will. But first, let's briefly discuss the increasing popularity of multi-GPU environments.

The popularity of multi-GPU environments

Going back 10 years ago, it was inconceivable to think of a machine with more than one GPU. Besides the high cost of this device, the applicability of a GPU was restricted to solving scientific computing problems, which is a niche that's exploited only by universities and research institutes. However, with the boom of **artificial intelligence** (**AI**) workloads, the usage of GPU was tremendously democratized across all sorts of companies.

Moreover, with the massive adoption of cloud computing in the last few years, we started to have cloud providers offering multi-GPU instances at a competitive price. In **Amazon Web Services** (**AWS**), for example, there are a variety of instances endowed with multiple GPUs such as **p5.48xlarge**, **p4d.24xlarge**, and **p3dn.24xlarge**, which provides 8 NVIDIA GPUs for the H100, A100, and V100 models, respectively.

Microsoft Azure and **Google Cloud Platform** (**GCP**) also have multi-GPU instances. The former offers the **NC96ads** with 4 NVIDIA A100s, while the latter offers the **a3-highgpu-8g** instance equipped with 8 NVIDIA H100s. Even second-tier cloud providers such as IBM, Alibaba, and **Oracle Cloud Infrastructure** (**OCI**) have multi-GPU instances.

Looking at the on-premises side, we have important vendors such as Supermicro, HP, and Dell offering multi-GPU platforms in their portfolio. NVIDIA, for example, provides a fully integrated server specially designed to run AI workloads known as the DGX system. DGX version 1, for example, is equipped with 8 GPUs of Volta or Pascal architecture, while DGX version 2 has twice the number of GPUs of its predecessor.

Considering the increasing popularity of these environments, it is more than reasonable to say that data scientists and machine learning engineers will have access to these platforms sooner or later. Note that many professionals already have these environments in their hands, though they do not know how to exploit them.

> **Note**
> Although a multi-GPU environment can provide an outstanding performance improvement for the training process, it has some drawbacks, such as the high cost to acquire and maintain such environments, and the huge amount of energy needed to control the temperature of these devices.

To use this resource efficiently, we must learn the fundamental characteristics of this environment. So, let's take our first step in that direction and understand how GPUs are connected to this platform.

Understanding multi-GPU interconnection

A multi-GPU environment can be seen as a pool of resources where different users can allocate devices individually to execute their training process. However, in the context of distributed training, we are interested in using more than one device simultaneously – that is, we will use each GPU to run a model replica of the distributed training process.

As the gradient that's obtained by each model replica must be shared among all other replicas, the GPUs in a multi-GPU environment must be connected so that the data can flow across the multiple devices available on the system. There are three types of GPU connection technologies: PCI Express, NVLink, and NVSwitch.

> **Note**
>
> You can find a comparison between these technologies in the paper entitled *Evaluating Modern GPU Interconnect: PCIe, NVLink, NV-SLI, NVSwitch and GPUDirect*, by Ang Li and others. You can access this paper at https://ieeexplore.ieee.org/document/8763922.

The following sections describe each of them.

PCI Express

PCI Express, also known as PCIe, is the default bus to connect all sorts of devices, such as network cards, disks, and GPUs, to the computer system, as shown in *Figure 10.1*. Therefore, PCIe is not a particular technology to interconnect GPUs. Au contraire, PCIe is a general and vendor-agnostic expansion bus that connects peripherals to the system, including GPUs:

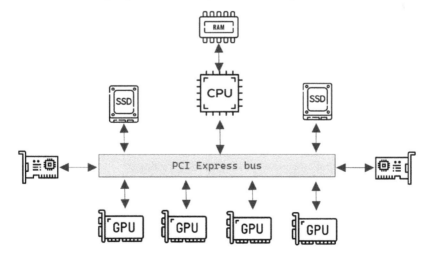

Figure 10.1 – PCIe interconnection technology

PCIe interconnects peripherals through two main components: the **PCIe root complex** and the **PCIe switch**. The former connects the entire PCIe subsystem to the CPU, while the latter is used to connect endpoint devices (peripherals) to the subsystem.

> **Note**
>
> The PCIe root complex is also known as the PCIe host bridge or PHB. In modern processors, the PCIe host bridge is placed inside the CPU.

As shown in *Figure 10.2*, PCIe uses switches to organize the subsystem on a hierarchical basis, where devices connected to a common switch belong to the same hierarchical level. Peripherals at the same hierarchical level communicate with each other at a lower cost than those in distinct levels of the hierarchical structure:

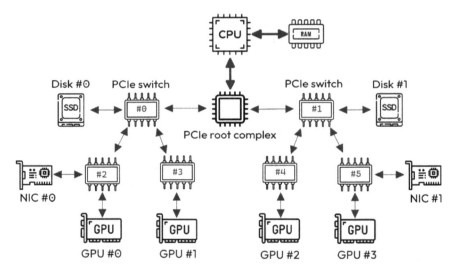

Figure 10.2 – PCIe subsystem

For example, the communication between GPU #0 and NIC #0 is faster than it is between GPU #1 and NIC #0. This happens because the first tuple is connected to the same switch (switch #2), while the devices of the last tuple are connected to different switches.

Similarly, the communication between GPU #3 and Disk #1 is cheaper than between GPU #3 and Disk #0. In the latter case, GPU #3 should traverse three switches and the root complex to reach Disk #0, whereas Disk #1 is far from GPU #3 by only two switches.

PCI Express does not provide a way to connect one GPU to another directly or to connect all GPUs. To overcome this issue, NVIDIA has invented a new interconnection technology called NVLink, as described in the next section.

NVLink

NVLink is an NVIDIA proprietary interconnection technology that allows us to connect pairs of GPUs directly to each other. NVLink provides superior data transfer rates compared to PCIe. A single NVLink can provide a data transfer of 25 GB per second, while PCIe allows a maximum data transfer rate of 1 GB per second.

Modern GPU architectures support more than one NVLink connection. Each link can be used to connect the GPU to different GPUs (as shown in *Figure 10.3 (a)*) or to bond the links to increase the bandwidth between two or more GPUs (as shown in *Figure 10.3 (b)*). The P100 and V100 GPUs, for example, support four and six NVLink connections, respectively:

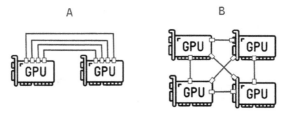

Figure 10.3 – NVLink connections

Nowadays, NVLink is the best option to interconnect NVIDIA GPUs. The benefits of using NVLink rather than PCIe are quite obvious. With NVLink, we can connect GPUs directly, reducing latency and improving bandwidth.

Notwithstanding, PCIe overcomes NVLink in one aspect: *scalability*. Due to the finite number of connections present in GPU, NVLink will not be able to connect a certain number of devices altogether. For example, it is impossible to connect eight GPUs altogether if each GPU supports only four NVLink connections. On the other hand, PCIe can connect any number of devices through PCIe switches.

To surpass this scalability problem, NVIDIA has created a complementary technology for NVLink known as **NVSwitch**. We'll learn about it in the next section.

NVSwitch

NVSwitch extends the degree of connectivity of GPUs by using NVLink switches. Roughly speaking, the idea behind NVSwitch is similar to the usage of switches on PCIe technology – that is, both interconnections rely on components acting like a concentrator or a hub. This component is used to link and aggregate devices:

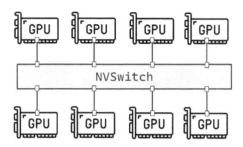

Figure 10.4 – NVSwitch interconnection topology

As shown in *Figure 10.4*, we can use NVSwitch to connect eight GPUs, regardless of the number of NVLinks supported by each GPU. Other arrangements involve NVLink and NVSwitch, such as the one presented in *Figure 10.5*:

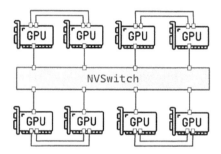

Figure 10.5 – Example of a topology using NVLink and NVSwitch

In the example illustrated in *Figure 10.5*, all GPUs are connected to themselves through NVSwitch. However, some pairs of GPUs are connected with two NVLinks, thus doubling the data transfer rate between these pairs. It is also possible to use more than one NVSwitch to provide total connectivity of GPUs, besides improving the connection among pairs or tuples of devices.

In summary, GPUs can be connected through distinct communication technologies that provide different data transfer rates and distinct ways to connect devices. As a result, we can have more than one path to connect two or more devices in a multi-GPU environment.

The way devices are connected in a system is called **interconnection topology** and is something that plays a vital role in the performance optimization of the training process. Let's jump to the next section to understand why the topology is worthy of our attention.

How does interconnection topology affect performance?

To understand the impact of interconnection topology on training performance, let's consider an analogy. Consider a city with multiple roads, such as freeways, highways, and streets, where each type of road has characteristics related to speed limit, congestion, and so forth. As the city has many roads, we have distinct ways to reach the same destination. Therefore, we need to decide which path is the best one to make our route as fast as possible.

We can think of the interconnection topology as the city described in our analogy. In the city, communication between devices can take distinct paths, where some paths are fast, such as the highways, and other ones are slow, such as a regular street. As stated in the city analogy, we should always choose the fastest connection between devices being used in the training process.

To have an idea of the potential impact of an unaware topology selection of devices, consider the block diagram illustrated in *Figure 10.6*, which represents an environment that's used to run highly intensive computing workloads as the training process:

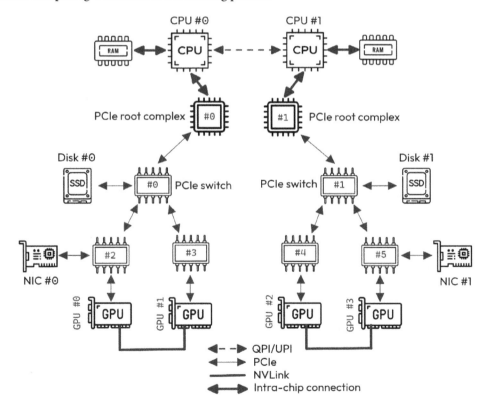

Figure 10.6 – Example of a system interconnection diagram

> **Note**
> The diagram shown in *Figure 10.6* is a simplified version of a real interconnection topology. Hence, we should consider it a didactic representation of a real topology scheme.

The environment illustrated in *Figure 10.6* can be classified as a multi-device platform because it has multiple GPUs, CPUs, and other important components, such as ultra-fast disks and network cards. Alongside multiple devices, such platforms also employ multiple interconnection technologies, such as the ones we learned about in the previous section.

Supposing we intend to use two GPUs to execute the distributed training process on the system described in *Figure 10.6*, which ones should we pick up?

If we choose GPU #0 and GPU #1, the communication will be fast since these devices are connected through NVLink connections. On the other hand, if we select GPU #0 and GPU#3, the communication will traverse the entire PCIe subsystem. Besides having a lower bandwidth than NVLink, communication through PCIe in this scenario will cross various PCIe switches, two PCIe root complexes, and both CPUs.

Naturally, we must choose the option that delivers the best communication performance, which can be achieved by using links with higher data transfer rates and using the nearest devices. In other words, *we need to use GPUs with the highest affinity*.

You might be wondering how to discover the interconnection topology of your environment. We'll learn how to do this in the next section.

Discovering the interconnection topology

To discover the interconnection topology of NVIDIA GPUs, we just need to execute the nvidia-smi command with two parameters:

```
maicon@packt:~$ nvidia-smi topo -m
```

The topo parameter stands for topology and provides options to get more information about the interconnection topology adopted in the system. The -m option tells nvidia-smi to print the GPU affinity in a matrix format.

The matrix that's printed by nvidia-smi reveals the affinity between each possible pair of GPUs available in the system. As the affinity between the same device is illogical, the matrix diagonal is marked with X. In the remaining coordinates, the matrix exhibits a label to denote the best connection type available for that pair of devices. The possible labels for the matrix are as follows (adapted from the nvidia-smi manual):

- **SYS**: The connection traversing PCIe as well as the interconnection between NUMA nodes (for example, QPI/UPI interconnections)

- **NODE**: The connection traversing PCIe as well as the interconnection between the PCIe root complex within a NUMA node

- **PHB**: The connection traversing PCIe as well as a PCIe root complex (PCIe host bridge)

- **PXB**: The connection traversing multiple PCIe bridges (without traversing any PCIe root complex)

- **PIX**: The connection traversing, at most, a single PCIe bridge

- **NV#**: The connection traversing a bonded set of # NVLinks

Let's evaluate an example of an affinity matrix generated by nvidia-smi. The matrix illustrated in *Table 10.1* was generated from an environment comprised of 8 GPUs:

	GPU0	GPU1	GPU2	GPU3	GPU4	GPU5	GPU6	GPU7
GPU0	X	NV1	NV1	NV2	NV2	SYS	SYS	SYS
GPU1	NV1	X	NV2	NV1	SYS	NV2	SYS	SYS
GPU2	NV1	NV2	X	NV2	SYS	SYS	NV1	SYS
GPU3	NV2	NV1	NV2	X	SYS	SYS	SYS	NV1
GPU4	NV2	SYS	SYS	SYS	X	NV1	NV1	NV2
GPU5	SYS	NV2	SYS	SYS	NV1	X	NV2	NV1
GPU6	SYS	SYS	NV1	SYS	NV1	NV2	X	NV2
GPU7	SYS	SYS	SYS	NV1	NV2	NV1	NV2	X

Table 10.1 – Example of an affinity matrix generated by nvidia-smi

The affinity matrix described in *Table 10.1* tells us that some GPUs are connected through two NVLinks (labeled NV2), while other ones are connected with just one NVLink (labeled NV1). In addition, many other GPUs do not share an NVLink connection, being connected only by the largest path in the system (labeled SYS).

So, if we needed to select two GPUs to work together in the distributed training process, it would be recommended to use, for example, GPUs #0 and #3, #0 and #4, and #1 and #2 because these pairs of devices are connected by two bonded NVLinks. Conversely, the worse option would be using GPUs #0 and #5 or #2 and #4 since the connection between these devices crosses the entire system.

If we are interested in knowing the affinity of two specific devices, we can execute nvidia-smi with the -i parameter, followed by the GPU's ID:

```
maicon@packt:~$ nvidia-smi topo -p -i 0,1
Device 0 is connected to device 1 by way of multiple PCIe switches.
```

In this example, GPUs #0 and #1 are connected through multiple PCIe switches, though they do not traverse any PCIe root complex.

> **Note**
>
> Another way to map the topology of NVIDIA GPUs is using **NVIDIA Topology-Aware GPU Selection (NVTAGS)**. NVTAGS is a toolset created by NVIDIA to automatically determine the fastest communication channel between GPUs. For more information about NVTAGS, you can access this link: https://developer.nvidia.com/nvidia-nvtags

Setting GPU affinity

The easiest way to set GPU affinity is by using the CUDA_VISIBLE_DEVICES environment variable. This variable allows us to indicate which GPUs will be visible to CUDA-based programs. To do this, we just need to specify the number IDs of the GPUs, separated by commas.

For example, considering an environment endowed with 8 GPUs, we must set CUDA_VISIBLE_DEVICES to a value of 2, 3 so that it can use GPUs #2 and #3:

```
CUDA_VISIBLE_DEVICES = "2,3"
```

Note that CUDA_VISIBLE_DEVICES defines which GPUs will be used by the CUDA program and not the number of devices. So, if the variable is set to 5, for example, the CUDA program will see only GPU device #5 and not five of the eight devices available in the system.

There are three ways to set CUDA_VISIBLE_DEVICES to select the GPUs we want to use in our training process:

1. **Exporting** the variable before starting the training program:

    ```
    maicon@packt:~$ export CUDA_VISIBLE_DEVICES="4,6"
    maicon@packt:~$ python training_program.py
    ```

2. **Setting** up the variable inside the training program:

    ```
    os.environ['CUDA_VISIBLE_DEVICES'] ="4,6"
    ```

3. **Defining** the variable in the same command line of the training program:

    ```
    maicon@packt:~$ CUDA_VISIBLE_DEVICES="4,6" python training_
    program.py
    ```

In the next section, we will learn how to code and launch distributed training on multiple GPUs.

Implementing distributed training on multiple GPUs

In this section, we'll show you how to implement and run distributed training on multiple GPUs using NCCL, the *de facto* communication backend for NVIDIA GPUs. We'll start by providing a brief overview of NCCL, after which we will learn how to code and launch distributed training in a multi-GPU environment.

The NCCL communication backend

NCCL stands for NVIDIA Collective Communications Library. As its name suggests, NCCL is a library that provides optimized collective operations for NVIDIA GPUs. Therefore, we can use NCCL to execute collective routines such as broadcast, reduce, and the so-called all-reduce operation. Roughly speaking, NCCL plays the same role as oneCCL does for Intel CPUs.

PyTorch supports NCCL natively, which means that the default installation of PyTorch for NVIDIA GPUs already comes with a built-in NCCL version. NCCL works on single or multiple machines and supports the usage of high-performance networks such as InfiniBand.

Along the lines of oneCCL and OpenMP, the behavior of NCCL can also be controlled through environment variables. For example, we can control the logging level of NCCL through the NCCL_DEBUG environment variable, which accepts the `trace`, `info`, and `warn` values. In addition, it is possible to filter the logs according to the subsystem by setting the NCCL_DEBUG_SUBSYS variable.

> **Note**
>
> The complete set of NCCL environment variables can be found at `https://docs.nvidia.com/deeplearning/nccl/user-guide/docs/env.html`.

In the next section, we will learn how to use NCCL as the communication backend in the distributed training process with multiple GPUs.

Coding and launching distributed training with multiple GPUs

The code and launching script to distribute the training process among multiple GPUs is almost the same as what was presented in *Chapter 9, Training with Multiple CPUs*. Here, we will learn how to adapt them for distributed training in a multi-GPU environment.

Coding the distributed training for multi-GPU

We only need to make two modifications to the multi-CPU code.

> **Note**
>
> The complete code shown in this section is available at https://github.com/
> PacktPublishing/Accelerate-Model-Training-with-PyTorch-2.X/blob/
> main/code/chapter10/nccl_distributed-efficientnet_cifar10.py.

The first modification concerns passing nccl as input for the backend parameter of the init_process_group method (line 77):

```
dist.init_process_group(backend="nccl", init_method="env://")
```

The second modification is the most important one, though. As we are running the training process in a multi-GPU environment, we need to guarantee that each process exclusively allocates one of the GPUs available on the system.

By doing this, we can utilize the process rank to define which device will be allocated to the process. For example, considering a muti-GPU environment comprised of four GPUs, process rank 0 will use GPU #0, process rank 1 will use GPU #1, and so on.

Although this change is essential to execute distributed training correctly, it is pretty simple to implement. We just need to attribute the process rank – stored in the my_rank variable – to the device variable:

```
device = my_rank
```

Concerning the GPU's affinity, you might be wondering, **how could we select GPUs we intend to use if each process allocates the GPU that corresponds to its rank?** This question is fair and usually leads to a lot of confusion. Fortunately, the answer is simple.

It turns out that the CUDA_VISIBLE_DEVICES variable abstracts the real GPU identification from the training program. So, if we set the variable to 6, 7, the training program will see only two devices – that is, devices identified with numbers 0 and 1. Thus, the processes with ranks 0 and 1 will allocate the GPU numbers 0 and 1, which are the real IDs of 6 and 7, respectively.

To summarize, only these two modifications are enough to get a piece of code ready to be executed in a multi-GPU environment. So, let's move on to the next step: launching the distributed training process.

Launching the distributed training process on a multi-GPU

The script to execute distributed training on multiple GPUs follows the same logic as the one we used to run distributed training on multiple CPUs:

```
TRAINING_SCRIPT=$1
NGPU=$2
TORCHRUN_COMMAND="torchrun --nnodes 1 --nproc-per-node $NGPU --master-
addr localhost $TRAINING_SCRIPT"
$TORCHRUN_COMMAND
```

In the GPU version, we pass the number of GPUs as an input parameter instead of the number of processes. Because we usually assign an entire GPU to a single process, the number of processes in distributed training is equal to the number of GPUs we intend to use.

Concerning the command line to execute the script, there is no difference between the CPU and GPU versions. We just need to call the script's name and inform the training script, followed by the number of GPUs:

```
maicon@packt:~$ ./launch_multiple_gpu.sh nccl_distributed-
efficientnet_cifar10.py 8
```

We can also adapt the script so that it uses containers, as we did with the CPU implementation:

```
TRAINING_SCRIPT=$1
NGPU=$2
SIF_IMAGE=$3
TORCHRUN_COMMAND="torchrun --nnodes 1 --nproc-per-node $NGPU --master-
addr localhost $TRAINING_SCRIPT"
apptainer exec --nv $SIF_IMAGE $TORCHRUN_COMMAND
```

The unique difference in the GPU implementation concerns the Apptainer command line. When using NVIDIA GPUs, we need to call Apptainer with the --nv parameter to enable support for these devices within the container.

> **Note**
> The complete code shown in this section is available at https://github.com/
> PacktPublishing/Accelerate-Model-Training-with-PyTorch-2.X/
> blob/main/scripts/chapter10/launch_multiple_gpu.sh.

Now, let's see how fast distributed training with multiple GPUs can be.

Experimental evaluation

To evaluate distributed training on multiple GPUs, we have trained the EfficientNet model against the CIFAR-10 dataset over 25 epochs by using a single machine equipped with 8 NVIDIA A100 GPUs. As a baseline, we will use the execution time of training this model with only 1 GPU, which was equal to 707 seconds.

The execution time of training the model with 8 GPUs was equal to 109 seconds, representing an impressive performance improvement of 548% compared to the execution time spent to train the model with just 1 GPU. In other words, the distributed training with 8 GPUs was almost 6.5 times faster than the singular training approach.

Nevertheless, as happened in the distributed training process with multiple CPUs, the model accuracy was also penalized by the distributed training process running multiple GPUs. With only 1 GPU, the trained model achieved an accuracy of 78.76%, but with 8 GPUs, the accuracy decreased to 68.82%.

This difference in model accuracy is relevant; therefore, we should not put it aside. Au contraire, we should take it into account when distributing the training process among multiple GPUs. For example, if we cannot tolerate a 10% difference in model accuracy, we should try reducing the number of GPUs.

To give you an idea of the relationship between performance gain and the corresponding model accuracy, we conducted additional tests. The results are shown in *Table 10.2*:

Number of GPUs	Execution Time	Accuracy
1	707	78.76%
2	393	74.82%
3	276	72.70%
4	208	70.72%
5	172	68.34%
6	142	69.44%
7	122	69.00%
8	109	68.82%

Table 10.2 – Results of distributed training with multiple GPUs

As shown in *Table 10.2*, the accuracy tends to decrease as the number of GPUs increases. However, if we take a closer look, we'll realize that 4 GPUs achieve a very good performance improvement (240%) while keeping the accuracy above 70%.

It is also interesting to note that model accuracy decreased by 4% when we used 2 GPUs in the training process. This result shows that the distributed training impacts accuracy, even using the smallest possible number of GPUs.

On the other hand, the model accuracy remained almost stable at around 68% from 5 devices onward, though performance improvement kept rising.

In short, it is essential to pay attention to model accuracy when increasing the number of GPUs in the distributed training process. Otherwise, a blind pursuit of performance improvement can lead to an undesirable result in the training process.

The next section provides a couple of questions to help you retain what you have learned in this chapter.

Quiz time!

Let's review what we have learned in this chapter by answering a few questions. Initially, try to answer these questions without consulting the material.

> **Note**
>
> The answers to all these questions are available at `https://github.com/ PacktPublishing/Accelerate-Model-Training-with-PyTorch-2.X/ blob/main/quiz/chapter10-answers.md`.

Before starting the quiz, remember that this is not a test! This section aims to complement your learning process by revising and consolidating the content covered in this chapter.

Choose the correct option for the following questions.

1. Which are the three main types of GPU interconnection technologies?

    A. PCI Express, NCCL, and GPU-Link.

    B. PCI Express, NVLink, and NVSwitch.

    C. PCI Express, NCCL, and GPU-Switch.

    D. PCI Express, NVML, and NVLink.

2. NVLink is a proprietary interconnection technology that allows you to do which of the following?

    A. Connect the GPU to the CPU.

    B. Connect the GPU to the main memory.

    C. Connect pairs of GPUs directly to each other.

    D. Connect the GPU to the network adapter.

3. Which environment variable is used to define GPU affinity?

    A. `CUDA_VISIBLE_DEVICES`.

    B. `GPU_VISIBLE_DEVICES`.

    C. `GPU_ACTIVE_DEVICES`.

    D. `CUDA_AFFINITY_DEVICES`.

4. What is NCCL?

    A. NCCL is an interconnection technology that's used to link NVIDIA GPUs.

    B. NCCL is a library that's used to profile programs running on NVIDIA GPUs.

C.   NCCL is a compiler toolkit that's used to generate optimized code for NVIDIA GPUs.

D.   NCCL is a library that provides optimized collective operations for NVIDIA GPUs.

5.   Which program launcher can be used to run distributed training on multiple GPUs?

A.   GPUrun.

B.   Torchrun.

C.   NCCLrun.

D.   oneCCL.

6.   If we set the CUDA_VISIBLE_DEVICES environment variable to a value of "2 , 3", which device numbers will be passed to the training script?

A.   2 and 3.

B.   3 and 2.

C.   0 and 1.

D.   0 and 7.

7.   How can we obtain more information about the interconnection topology that's adopted in a given multi-GPU environment?

A.   Running the nvidia-topo-ls command with the -interconnection option.

B.   Running the nvidia-topo-ls command with the -gpus option.

C.   Running the nvidia-smi command with the -interconnect option.

D.   Running the nvidia-smi command with the -topo option.

8.   Which component is used by the PCI Express technology to interconnect PCI Express devices in a computing system?

A.   PCIe switch.

B.   PCIe nvswitch.

C.   PCIe link.

D.   PCIe network.

Summary

In this chapter, we learned how to distribute the training process across multiple GPUs by using NCCL, the optimized NVIDIA library for collective communication.

We started this chapter by understanding how a multi-GPU environment employs distinct technologies to interconnect devices. Depending on the technology and interconnection topology, the communication between devices can slow down the entire distributed training process.

After being introduced to the multi-GPU environment, we learned how to code and launch distributed training on multiple GPUs by using NCCL as the communication backend and `torchrun` as the launch provider.

The experimental evaluation of our multi-GPU implementation showed that distributed training with 8 GPUs was 6.5 times faster than running with a single GPU; this is an expressive performance improvement. We also learned that model accuracy can be affected by performing distributed training on multiple GPUs, so we must take it into account when increasing the number of devices that are used in the distributed training process.

To end our journey of accelerating the training process with PyTorch, in the next chapter, we will learn how to distribute the training process among multiple machines.

11

Training with Multiple Machines

We've finally arrived at the last mile of our performance improvement journey. In this last stage, we will broaden our horizons and learn how to distribute the training process across multiple machines or servers. So, instead of using four or eight devices, we can use dozens or hundreds of computing resources to train our models.

An environment comprised of multiple connected servers is usually called a computing cluster or simply a cluster. Such environments are shared among multiple users and have technical particularities such as a high bandwidth and low latency network.

In this chapter, we'll describe the characteristics of computing clusters that are more relevant to the distributed training process. After that, we will learn how to distribute the training process among multiple machines using Open MPI as the launcher and NCCL as the communication backend.

Here is what you will learn as part of this chapter:

- The most relevant aspects of computing clusters
- How to distribute the training process among multiple servers
- How to use Open MPI as a launcher and NCCL as the communication backend

Technical requirements

You can find the complete code of examples mentioned in this chapter in the book's GitHub repository at https://github.com/PacktPublishing/Accelerate-Model-Training-with-PyTorch-2.X/blob/main.

You can access your favorite environments to execute this notebook, such as Google Colab or Kaggle.

What is a computing cluster?

A computing cluster is a system of powerful servers interconnected by a high-performance network, as shown in *Figure 11.1*. This environment can be provisioned on-premises or in the cloud:

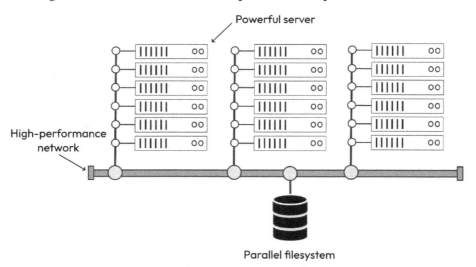

Figure 11.1 – A computing cluster

The computing power provided by these machines is combined to solve complex problems or to execute highly intensive computing tasks. A computing cluster is also known as a **high-performance computing (HPC)** system.

Each server has powerful computing resources such as multiple CPUs and GPUs, fast memory devices, ultra-fast disks, and special network adapters. Moreover, a computing cluster often has a parallel filesystem, which provides high transfer I/O rates.

Although not formally defined, we conventionally use the term "cluster" to reference environments comprised of four machines at least. Some computing clusters have a half-dozen machines, while others have more than two or three hundred servers.

Each task submitted to the cluster is called a **job**. When submitting a job, the user asks for a given number and type of resource and indicates which program should be executed in the environment. Therefore, any computing task running in the cluster is considered a job.

Jobs and operating system processes have many things in common. Like a process, a job is also identified by a unique number in the system, has a life cycle comprised of a finite set of states, and belongs to a system user.

As pictorially described in *Figure 11.2*, the bigger part of the servers is used as **computing nodes** – in other words, machines are used exclusively to run jobs. A couple of machines, called **management nodes**, are used to perform administrative tasks, such as monitoring and installation, or to provide auxiliary and complimentary services, such as a user access entering point, commonly called a **login node**.

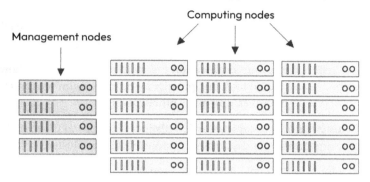

Figure 11.2 – Management and computing nodes

Another vital service hosted on management nodes is the cluster management system or **workload manager**. As the cluster is shared among multiple users, it is mandatory to have a workload manager to guarantee the fair and efficient usage of resources. Let's learn about it in the next section.

Workload manager

A workload manager is responsible for keeping the cluster environment running smoothly by providing fair and efficient usage of the resources. As illustrated in *Figure 11.3*, the workload manager is placed between users and resources to receive requests from the users, process these requests, and grant or deny access to the required resources:

Figure 11.3 – Workload manager

Among the tasks this system executes, two of them stand out from the others: resource management and job scheduling. The following sections briefly describe each of them.

Resource management

Roughly speaking, a cluster can be seen as a pool of shared resources where these resources are consumed by a set of users. The main goal of **resource management** concerns guaranteeing the fair usage of these resources.

By fair usage, we mean avoiding imbalance situations such as a greedy user consuming all the available resources, preventing less frequent users from getting access to the environment.

Resource management relies on a **resource allocation policy** to decide when and how to attend to users' requests. This policy can be used to define priority levels, maximum usage time, maximum number of running jobs, type of resources, and many other conditions.

From these policies, cluster administrators can assign distinct strategies to the users by following a criterion defined by the organization or department responsible for the cluster.

For example, a cluster administrator could define two resource allocation policies to limit issues such as the maximum number of running jobs, types of resources, and the maximum allowed time to run a job. As illustrated in *Figure 11.4*, the more restrictive policy, named **A**, could be applied to the group of users **X**, while the more permissive policy, named **B**, could be assigned to users **Y**:

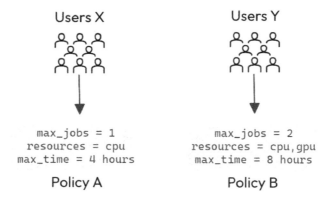

Figure 11.4 – Example of resource allocation policies

By doing this, the cluster administrator can determine distinct usage profiles for the cluster.

Jobs scheduling

A workload manager is also responsible for the efficient use of resources. To reach this goal, the workload manager must perform an optimal (or suboptimal) allocation of jobs on the computing nodes. This process is called **job scheduling** and is defined as the task of deciding where to put the new job to run.

As illustrated in *Figure 11.5*, the workload manager must select the computing node in which the new job will be executed. To make this decision, the workload manager evaluates the amount and type of requested resources and the number and type of available resources in all computing nodes. By doing this, the job scheduling gets a list of potential nodes suitable to execute the job – in other words, nodes with enough resources to satisfy the job requirements.

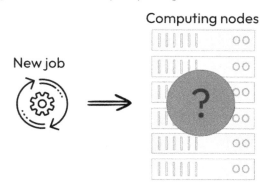

Figure 11.5 – Job scheduling

From the list of potential nodes, the job scheduling needs to decide which one will be chosen to execute the job. This decision is made according to a scheduling strategy that may prioritize fulfilling all nodes before using another one or spreading jobs to the computing nodes as much as possible to avoid interference caused by access contention to shared resources.

These sections have provided a general explanation of how a workload manager works. In practice, a real workload manager has particularities and ways to implement the resource management and job scheduling processes.

There are a couple of workload managers out there. Some are proprietary and vendor-specific, whereas others are free and open source, such as SLURM, the most widely used workload manager nowadays. Let's meet this system in the next section.

Meeting the SLURM Workload Manager

SLURM's website describes it as an *"open source, fault-tolerant, and highly scalable cluster management and job scheduling system for large and small Linux clusters"* – it is right.

> **Note**
> You can find more information about SLURM at this link: `https://slurm.schedmd.com/`

SLURM is powerful, robust, flexible, and simple to use and administrate. In addition to the basic functionalities found in any workload manager, SLURM offers special capabilities such as **QOS** (**quality of service**), accounting, database storage, and an **API** (**application programming interface**) that allows you to get information about the environment.

This workload manager uses the concept of **partition** to group computing nodes and to define resource allocation policies on the available resources, as shown in *Figure 11.6*:

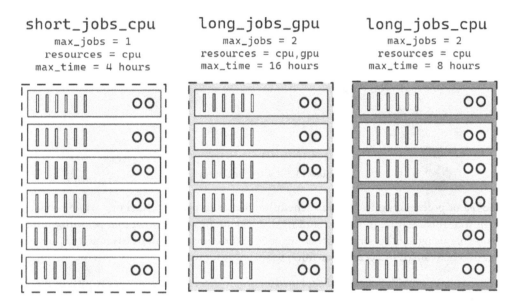

Figure 11.6 – Example of SLURM partitions

In the example depicted in *Figure 11.6*, we have three partitions, each with eight computing nodes but with distinct resource allocation policies. The short_jobs_cpu partition, for example, allows you to run a job for a maximum time of four hours, while the long_jobs_cpu partition has a maximum execution time of eight hours. Moreover, only the long_jobs_gpu partition has computing nodes that can run GPU jobs.

> **Note**
>
> SLURM uses the term **partition** to denote what other workload managers call a **queue**. Nevertheless, a partition works essentially as a queue, receiving job requests and organizing their execution concerning resource allocation and job scheduling policies.

Therefore, the partition is a central aspect of the SLURM architecture. With partitions, cluster administrators can employ distinct resource allocation policies, besides separating nodes to run specific applications or leaving them separated to be used exclusively by a department or group of users.

When users submit jobs, they must indicate the partition where the job will run. Otherwise, SLURM will submit the job to the default partition, which is defined by the cluster administrator.

When using a computing cluster, we can come across other workload managers such as OpenPBS, Torque, LSF, and HT Condor. However, due to its increased adoption in the HPC industry, it is more feasible to encounter SLURM as a workload manager on clusters you will have access to. So, we encourage you to invest some time to deepen your knowledge of SLURM.

Besides workload management, computing clusters have another component particularly important to these environments: the high-performance network. The next section provides a very brief explanation of this component.

Understanding the high-performance network

An essential difference between running distributed training on a single machine and using a computing cluster is the network used to interconnect the servers. The network imposes an additional bottleneck to the communication among the processes participating in distributed training. Fortunately, computing clusters usually have a high-performance network to connect all the servers in the environment.

This high-performance network differs from the regular ones because of its high bandwidth and very low latency. For example, the maximum theoretical bandwidth of Ethernet 10 Gbps is around 1.25 GB/s, whereas **NVIDIA InfiniBand** 100 Gbps EDR, which is one of the most adopted high-performance networks, provides a bandwidth near to 12.08 GB/s. In other words, a high-performance network can deliver 10 times more bandwidth than a regular network.

> **Note**
> You can find more information about NVIDIA InfiniBand at this link: `https://www.nvidia.com/en-us/networking/products/infiniband/`

Although the high bandwidth provided by InfiniBand is stunning, what makes a high-performance network so special is its very low latency. Compared with Ethernet 10 Gbps, the latency of InfiniBand 100 Gbps EDR can be almost four times lower. A low latency is crucial for the execution of distributed applications. As these applications exchange many messages during the computation, a single delay in the messages can throttle the entire application.

Besides the high bandwidth and low latency, high-performance networks (including InfiniBand) possess another special functionality named **remote direct memory access** or **RDMA**. Let's learn about it in the next section.

RDMA

RDMA is a functionality provided by high-performance networks to reduce communication latency among devices. Before understanding the advantages of using RDMA, we should first remember how regular communications work under the hood.

A **regular data transmission** involving two GPUs follows the procedure illustrated in *Figure 11.7*:

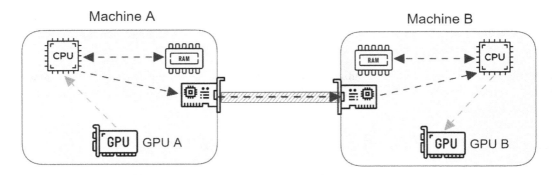

Figure 11.7 – Regular data transmission between two GPUs

First, **GPU A** asks the CPU to send data to **GPU B**, which is located in **Machine B**. The CPU receives the request and creates a buffer on the main memory of **Machine A** to store the data to be transmitted. Next, **GPU A** sends the data to the main memory and notifies the CPU that the data is already available on the main memory. So, the CPU on **Machine A** copies the data from the main memory to the network adapter's buffer. Then, the network adapter on **Machine A** establishes the communication channel with **Machine B** and sends the data. Finally, the network adapter on **Machine B** receives the data, and **Machine B** executes the same steps previously performed by **Machine A** to deliver the received data to **GPU B**.

Notice that this procedure involves many intermediary copies of the data on the main memory; in other words, the data is copied from the GPU's memory to the main memory and from the main memory to the network adapter's buffer, in both directions. So, it is easy to see that this procedure imposes a high overhead on the communication between GPUs located on remote machines.

To overcome this problem, applications can transfer data between devices by using RDMA. As shown in *Figure 11.8*, RDMA can transfer data directly from one GPU to another by using a high-performance network. After completing an initial setup, network adapters and GPUs become able to *transfer data without involving the CPU and main memory*. As a consequence, RDMA eliminates a bunch of intermediary copies of the transmission data, thus lowering the communication latency vastly. This is the reason why RDMA is also known as **zero-copy** transmission.

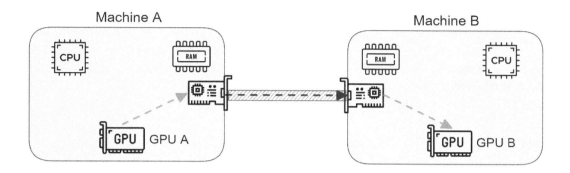

Figure 11.8 – RDMA between two GPUs

To use RDMA, the high-performance network, devices, and operating system must support this capability. So, if we intend to use this resource, we should first verify with the cluster administrator whether this resource is available and how to use it in the environment.

After learning the main characteristics of the computing cluster environment, we can move on and learn how to implement distributed training on multiple machines.

Implementing distributed training on multiple machines

This section shows how to implement and run the distributed training on multiple machines by using Open MPI as the launch provider and NCCL as the communication backend. Let's start by introducing Open MPI.

Getting introduced to Open MPI

MPI stands for **message passing interface** and is a standard that specifies a set of communication routines, data types, events, and operations used to implement distributed memory-based applications. MPI is so relevant to the HPC industry that it is ruled and maintained by a forum comprised of distinguished scientists, researchers, and professionals around the globe.

> **Note**
>
> You can find more information about MPI at this link: `https://www.mpi-forum.org/`

Therefore, MPI, strictly speaking, is not software; it is a standard specification that can be used to implement a software, tool, or library. Like non-proprietary programming languages such as C and Python, MPI also has many implementations. Some of them are vendor-specific, such as Intel MPI, while others are free and open source, such as MPICH.

Among all implementations, **Open MPI** sticks out as one of the most known and adopted implementations of MPI. Open MPI is free, open source, and maintained by a consortium composed of many major tech players such as AMD, AWS, IBM, Intel, and NVIDIA. The consortium also counts on renowned universities and research institutes such as the Los Alamos National Laboratory and Inria, the French National Institute for Research in Computer Science and Control.

> **Note**
> You can find more information about Open MPI at this link: `https://www.open-mpi.org/`

Open MPI is more than just a library to implement the MPI routines on applications. It is a toolset that provides other components such as compilers, debuggers, and a complete runtime mechanism.

The following section presents how to execute an Open MPI program. This knowledge is relevant to learning how to launch the distributed training process.

Executing an Open MPI program

To execute an Open MPI program, we should call the `mpirun` command and pass the MPI program and number of processes as parameters:

```
maicon@packt:~$ mpirun --np 2 my_mpi_program
```

The `--np` parameter tells Open MPI the number of processes it must create. If nothing else is informed to Open MPI, it will create these processes locally – in other words, in the machine in which the `mpirun` command was called. To instantiate processes in remote machines, we must use the `--host` argument followed by the list of remote machines separated by commas:

```
maicon@packt:~$ mpirun --np 2 --host r1:1,r2:1 my_mpi_program
```

In the previous example, `mpirun` will execute two processes, one on the `r1` remote machine and the other one on the `r2` remote machine. The value put after the name of the remote machines indicates the number of slots (or processes) the machine is willing to take. For example, if we want to execute six processes, four in the `r1` remote machine and two on the `r2` remote machine, we should call the following `mpirun` command:

```
maicon@packt:~$ mpirun --np 6 --host r1:4,r2:2 my_mpi_program
```

Open MPI sets some environment variables to the scope of each process created by the `mpirun` command. These environment variables give essential information about the distributed environment, such as the process's rank. Three of them are particularly interesting to our case:

- `OMPI_COMM_WORLD_SIZE`: the total number of processes participating in the distributed execution

- `OMPI_COMM_WORLD_RANK`: the **global rank** of the process

- `OMPI_COMM_WORLD_LOCAL_RANK`: the **local rank** of the process

To understand the difference between the global and local ranks, let's take the previous example of running six processes and put down the values of global and local ranks for each process in *Table 11.1*:

Process	Remote Machine	Global Rank	Local Rank
0	r1	0	0
1	r1	1	1
2	r1	2	2
3	r1	3	3
4	r2	4	0
5	r2	5	1

Table 11.1 – Global rank versus local rank

As shown in *Table 11.1*, the **global rank** is the global identification of the process – in other words, the rank of the process regardless of the machine it is running on. We can see it as a global identification of the process.

The **local rank** is the direct opposite; it identifies the process considering the machine it is running upon – in other words, the distributed environment boils down to that machine. For example, as we have two processes executing on the r2 machine, then the local rank of processes 5 and 6 are equal to 0 and 1, respectively.

The concept of local rank may seem counterintuitive and useless, but it is not. Local rank is very useful in distributed programs and especially convenient for our distributed training process. Wait and see!

Why use Open MPI and NCCL?

You may wonder why we use Open MPI as the launcher and NCCL as the communication backend. Indeed, maybe you are asking yourself the following questions:

1. Is it possible to use Open MPI as both the launcher and communication backend?

2. Is it possible to use NCCL as the communication backend and torchrun as the launcher?

The short answer to these questions is "*Yes, it is possible.*" However, there are some disadvantages to adopting these approaches. Let's discuss each of them.

As we are running the distributed training with multiple GPUs, the best communication backend for this case is surely the NCCL. Although it is possible to use Open MPI as the communication backend for this scenario, the collective operations provided by NCCL are the most optimized ones for NVIDIA GPUs.

So, now we know why we should choose NCCL rather than Open MPI as the communication backend. But why not use `torchrun` as the launch provider, as we have done so far?

Well, `torchrun` is an excellent choice to run the distributed training locally. However, to run the distributed training on multiple machines, we will need to execute a `torchrun` instance manually on each remote machine participating in the distributed environment.

Unlike `torchrun`, Open MPI natively supports the execution on remote machines more easily and elegantly. By using its runtime mechanism, Open MPI can smoothly create processes on remote machines, making our lives easier.

In short, we decided to use NCCL and Open MPI to get the best of the two worlds together.

Coding and launching the distributed training on multiple machines

The code to distribute the training process among multiple machines is almost the same as the one presented in *Chapter 10, Training with Multiple GPUs*. After all, we are going to execute multi-GPU training, but using multiple machines instead. Therefore, we are going to adapt the multi-GPU implementation to execute on multiple machines by using Open MPI as the launch provider.

Because we will use Open MPI as the launcher, the script used to launch the distributed training will not execute the `torchrun` command as we have done in the last two chapters. Thus, we will need to create a script from scratch to adopt Open MPI as the launching method.

Let's move to the following sections to learn how to adapt the multi-GPU implementation and create a launch script for the distributed training in a computing cluster environment.

Coding the distributed training for multiple machines

Compared to the muti-GPU implementation presented in *Chapter 10, Training with Multiple GPUs*, the code to run a distributed training in a computing cluster has the following three modifications:

```
os.environ['RANK'] = os.environ['OMPI_COMM_WORLD_RANK']
os.environ['WORLD_SIZE'] = os.environ['OMPI_COMM_WORLD_SIZE']
device = int(os.environ['OMPI_COMM_WORLD_LOCAL_RANK'])
```

> **Note**
>
> The complete code shown in this section is available at `https://github.com/PacktPublishing/Accelerate-Model-Training-with-PyTorch-2.X/blob/main/code/chapter11/nccl_mpi_distributed-efficientnet_cifar10.py`.

The first two modifications concern setting the environment variables, RANK and WORLD_SIZE, which are expected by the `init_process_group` method to create the communication group. As Open MPI uses other variable names to store this information, we need to explicitly define those variables in the code.

The third modification is related to defining the device (GPU, in this case) to be allocated to each process. As we have learned in the previous section, the local rank is an index that identifies processes running in each machine. Thus, we can use this information as an index to select the GPU used by each process. Therefore, the code must assign the content of the OMPI_COMM_WORLD_LOCAL_RANK environment variable to the `device` variable.

For example, consider the case of executing a distributed training comprised of eight processes by using two machines equipped with four GPUs each. The first four processes will have global and local ranks equal to 0, 1, 2, and 3. So, the process with a global rank of 0, which has a local rank of 0, will use the #0 GPU in the **first machine**, and so forth for the other processes in the first machine.

Concerning the four processes on the second machine, the process with a global rank of 4, which is the first process in the second machine, will have a local rank equal to 0. Thus, the process with a global rank of 4 will access the #0 GPU in the **second machine**.

Only these three modifications are enough to adjust the multi-GPU code to run on multiple machines. In the next section, let's find out how to launch the distributed training by using Open MPI.

Launching the distributed training on multiple machines

To use Open MPI as the launcher, we need to have it installed in the computing cluster environment. This installation should be provided by the cluster administrator since we are using Open MPI as an external component and not inside PyTorch. The cluster administrator should follow the installation instructions described on the Open MPI website.

Once we have Open MPI installed on the environment, we have two ways to launch the distributed training on multiple machines. We can execute it *manually* or *submit a job* to the workload manager. Let's first learn how to do it manually.

Manual execution

To execute the distributed training manually, we can use a launching script like this:

```
TRAINING_SCRIPT=$1
NPROCS= "16"
```

```
HOSTS="machine1:8,machine2:8"
COMMAND="python $TRAINING_SCRIPT"
export MASTER_ADDR="machine1"
export MASTER_PORT= "12345"
mpirun -x MASTER_ADDR -x MASTER_PORT --np $NPROCS --host $HOSTS
$COMMAND
```

> **Note**
>
> The complete code shown in this section is available at https://github.com/
> PacktPublishing/Accelerate-Model-Training-with-PyTorch-2.X/
> blob/main/scripts/chapter11/launch_multiple_machines.sh.

As we said before, this script differs from the one based on torchrun. Instead of calling the torchrun command, the script executes mpirun, as we have learned in previous sections. The mpirun command in this script is executed with five parameters. Let's take them one by one.

The first two parameters export the MASTER_ADDR and MASTER_PORT environment variables to the training program. This is done by using the -x parameter of mpirun.

By doing this, the init_process_group method can properly create the communication group. The MASTER_ADDR environment variable indicates the machine in which the launching script will be executed. In our case, it is executed in machine1. The MASTER_PORT environment variable defines the TCP port number used by the communication group to establish communication with all processes participating in the distributed environment. We can choose a higher number to avoid conflict with any bound TCP port.

The --np parameter determines the number of processes, and the --host parameter is used to indicate the list of machines in which mpirun will create processes. In this example, we are considering two machines named machine1 and machine2. Since each machine has eight GPUs, its name is followed by the number eight to indicate the maximum number of processes each server can execute.

The last parameter is the MPI-based program. In our case, we will pass the name of the Python interpreter followed by the name of the training script.

To execute this script for a program called distributed-training.py, we just need to run the following command:

```
maicon@packt:~$ ./launch_multiple_machines.sh distributed-training.py
```

> **Note**
>
> The complete code shown in this section is available at https://github.com/
> PacktPublishing/Accelerate-Model-Training-with-PyTorch-2.X/blob/
> main/scripts/chapter11/launch_multiple_machines_container.sh

Naturally, this script can be customized to accept other parameters such as the number of processes, the list of hosts, and so on. However, our intention here is to show the basic – though essential – way to manually execute the distributed training with Open MPI.

Job submission

Considering that the workload manager is SLURM, we must execute the following steps to submit a job to the computing cluster:

1. Create a batch script to submit the job to SLURM.
2. Submit the job with the sbatch command.

A batch script to submit a distributed training on SLURM will look like this:

```
#!/bin/bash
#SBATCH -n 16
#SBATCH --partition=long_job_gpu
#SBATCH --nodes=2
#SBATCH --gpus-per-node=8
export MASTER_ADDR=$(hostname)
export MASTER_PORT= "12345"
mpirun -x MASTER_ADDR -x MASTER_PORT --np 16 python /share/
distributed-training.py
```

This batch script will submit a job requesting two nodes and eight GPUs per node to be executed on the long_job_gpu partition. Like in the launching script, we also need to export the MASTER_ ADDR and MASTER_PORT variables so the init_process_group method can create the communication group.

After creating the script, we just need to submit the job by executing the following command:

```
maicon@packt:~$ sbatch distributed-training.sbatch
```

> **Note**
>
> The batch script presented before is just an illustrative example of how to submit a distributed training job on SLURM. As each computing cluster environment can have particularities, the best approach is always to take the guidelines from the cluster administrator concerning the usage of Open MPI. Anyway, you can consult the official SLURM documentation about running Open MPI jobs at https://slurm.schedmd.com/mpi_guide.html.

In the next section, we will look at the results of running the distributed training on two machines.

Experimental evaluation

To evaluate the distributed training on multiple machines, we have trained the EfficientNet model against the CIFAR-10 dataset over 25 epochs by using two machines, each one equipped with 8 GPUs NVIDIA A100. As a baseline, we will use the execution time of training this model with 8 GPUs in a single machine, which was equal to 109 seconds.

The execution time of training the model with 16 GPUs was equal to 64 seconds, representing a performance improvement of 70% compared to the execution time spent to train the model with eight GPUs in a single machine.

At first sight, this result can seem a little bit disappointing because we have used double the computing resources and got only a 70% performance improvement. As we used twice the number of resources, we should achieve 100% improvement.

However, we should remember that there is an additional component of this system: the interconnection between machines. Despite being a high-performance network, it is expected that an extra element has some impact on the performance. Even so, this result is quite good since we got closer to the maximum performance improvement we could achieve – in other words, 100%.

As expected, the model's accuracy decreased from 68.82% to 63.73%, corroborating the assertion about the relation between accuracy and the number of model replicas in the distributed training.

To summarize these results, we can highlight two interesting insights, as follows:

- We must always keep an eye on the model's quality when seeking performance improvement. As we have seen here and in the last two chapters, there is a potential depreciation of model accuracy in the face of an increased number of model replicas.

- We should ponder the possible impact caused by the interconnection network when deciding to distribute the training among multiple machines. Depending on the scenario, it could be more advantageous to keep the training inside a single machine with multiple GPUs rather than using multiple servers.

In short, a blind pursuit of performance improvement is often a bad idea because we can fall on resource wastage caused by a tiny performance improvement or a silent degradation of the model's quality. Therefore, we should always pay attention to the tradeoff between performance improvement, accuracy, and resource usage.

The next section provides a couple of questions to help you retain what you have learned in this chapter.

Quiz time!

Let's review what we have learned in this chapter by answering a few questions. First, try to answer these questions without consulting the material.

> **Note**
>
> The answers to all these questions are available at https://github.com/
> PacktPublishing/Accelerate-Model-Training-with-PyTorch-2.X/
> blob/main/quiz/chapter11-answers.md.

Before starting the quiz, remember that it is not a test at all! This section aims to complement your learning process by revising and consolidating the content covered in this chapter.

Choose the correct option for the following questions:

1. What is a task submitted to a computing cluster called?

    A. Thread.

    B. Process.

    C. Job.

    D. Work.

2. What are the main tasks executed by a workload manager?

    A. Resource management and job scheduling.

    B. Memory allocation and thread scheduling.

    C. GPU management and node scheduling.

    D. Resource management and node scheduling.

3. Which of the following is an open source, fault-tolerant, and highly scalable workload manager for large and small Linux clusters?

    A. MPI.

    B. SLURM.

    C. NCCL.

    D. Gloo.

4. A computing cluster is usually equipped with a high-performance network such as NVIDIA InfiniBand. Besides providing a high bandwidth, a high-performance interconnection provides which of the following?

    A. A high latency.

    B. A high number of connections.

    C. A low number of connections.

    D. A very low latency.

5. RDMA reduces drastically the communication latency between two remote GPUs because it enables which of the following?

    A. Allocation of higher memory space on GPUs.

    B. Special hardware capabilities on GPUs.

    C. Data transfer without involving the CPU and main memory.

    D. Data transfer without involving network adapters and switches.

6. Which of the following is the best definition of Open MPI?

    A. Open MPI is a compiler to create distributed applications.

    B. Open MPI is a toolset comprised of compilers, debuggers, and a complete runtime mechanism to create, debug, and run distributed applications.

    C. Open MPI is a standard that specifies a set of communication routines, data types, events, and operations used to implement distributed applications.

    D. Open MPI is a communication backend exclusively created to run the distributed training under PyTorch.

7. Consider the scenario in which a distributed training is running four processes under two machines (each machine is executing two processes). In this case, what are the ranks assigned by Open MPI for the two processes executing on the second machine?

    A. 0 and 1.

    B. 0 and 2.

    C. 2 and 3.

    D. 0 and 3.

8. Concerning the decision to distribute the training process among multiple machines or keep it in a single host, it is reasonable to ponder which of the following?

    A. The power consumption of using network adapters.

    B. The leak of memory space available on the network adapters.

    C. Nothing; it is always recommended to use multiple machines to run the distributed training.

    D. The impact the interconnection network may have on the communication between the processes participating in the distributed training.

Summary

In this chapter, we learned how to distribute the training process across multiple GPUs located on multiple machines. We used Open MPI as the launch provider and NCCL as the communication backend.

We decided to use Open MPI as the launcher because it provides an easy and elegant way to create distributed processes on remote machines. Although Open MPI can also be employed like the communication backend, it is preferable to adopt NCCL since it has the most optimized implementation of collective operations for NVIDIA GPUs.

Results showed that the distributed training with 16 GPUs on two machines was 70% faster than running with 8 GPUs on a single machine. The model accuracy decreased from 68.82% to 63.73%, which is expected since we have doubled the number of model replicas in the distributed training process.

This chapter ends our journey about learning how to accelerate the training process with PyTorch. More than knowing how to apply techniques and methods to speed up the model training, we expect that you have caught the foremost message of this book: performance improvement is not always related to new computing resources or novel hardware; it is possible to accelerate the training process with what we have on our hands by using the resources more efficiently.

Index

packtpub.com

Subscribe to our online digital library for full access to over 7,000 books and videos, as well as industry leading tools to help you plan your personal development and advance your career. For more information, please visit our website.

Why subscribe?

- Spend less time learning and more time coding with practical eBooks and Videos from over 4,000 industry professionals
- Improve your learning with Skill Plans built especially for you
- Get a free eBook or video every month
- Fully searchable for easy access to vital information
- Copy and paste, print, and bookmark content

Did you know that Packt offers eBook versions of every book published, with PDF and ePub files available? You can upgrade to the eBook version at packtpub.com and as a print book customer, you are entitled to a discount on the eBook copy. Get in touch with us at customercare@packtpub.com for more details.

At www.packtpub.com, you can also read a collection of free technical articles, sign up for a range of free newsletters, and receive exclusive discounts and offers on Packt books and eBooks.

Other Books You May Enjoy

If you enjoyed this book, you may be interested in these other books by Packt:

Mastering PyTorch

Ashish Ranjan Jha

ISBN: 978-1-80107-430-8

- Implement text, image, and music generating models using PyTorch
- Build a deep Q-network (DQN) model in PyTorch
- Deploy PyTorch models on mobiles and embedded devices
- Become well-versed with rapid prototyping using PyTorch with fast.ai
- Perform neural architecture search effectively using AutoML
- Easily interpret machine learning models using Captum
- Develop your own recommendation system using TorchRec
- Design ResNets, LSTMs, and graph neural networks
- Create language and vision transformer models using Hugging Face

Advanced Python Programming

Quan Nguyen

ISBN: 978-1-80181-401-0

- Write efficient numerical code with NumPy, pandas, and Xarray
- Use Cython and Numba to achieve native performance
- Find bottlenecks in your Python code using profilers
- Optimize your machine learning models with JAX
- Implement multithreaded, multiprocessing, and asynchronous programs
- Solve common problems in concurrent programming, such as deadlocks
- Tackle architecture challenges with design patterns

Packt is searching for authors like you

If you're interested in becoming an author for Packt, please visit `authors.packtpub.com` and apply today. We have worked with thousands of developers and tech professionals, just like you, to help them share their insight with the global tech community. You can make a general application, apply for a specific hot topic that we are recruiting an author for, or submit your own idea.

Share Your Thoughts

Now you've finished *Accelerate Model Training with PyTorch 2.X*, we'd love to hear your thoughts! Scan the QR code below to go straight to the Amazon review page for this book and share your feedback or leave a review on the site that you purchased it from.

`https://packt.link/r/1-805-12010-7`

Your review is important to us and the tech community and will help us make sure we're delivering excellent quality content.

Download a free PDF copy of this book

Thanks for purchasing this book!

Do you like to read on the go but are unable to carry your print books everywhere?

Is your eBook purchase not compatible with the device of your choice?

Don't worry, now with every Packt book you get a DRM-free PDF version of that book at no cost.

Read anywhere, any place, on any device. Search, copy, and paste code from your favorite technical books directly into your application.

The perks don't stop there, you can get exclusive access to discounts, newsletters, and great free content in your inbox daily

Follow these simple steps to get the benefits:

1. Scan the QR code or visit the link below

https://packt.link/free-ebook/978-1-80512-010-0

2. Submit your proof of purchase
3. That's it! We'll send your free PDF and other benefits to your email directly

www.ingramcontent.com/pod-product-compliance
Lightning Source LLC
Chambersburg PA
CBHW080523060326
40690CB00022B/5009